The Silver Dart

Skaters bring the *Silver Dart* into take-off position at Baddeck on February 24, 1909.

We dedicate this edition to

Mabel Bell

the woman of vision and the financier
of the Aerial Experiment Association—
whose determination helped make their success possible

The Silver Dart

The Story of the Hon. J.A.D. McCurdy Canada's First Pilot

with Photographs of the 1909 Flight of the *Silver Dart* and the Reconstructions and the Celebrations in 1959, 1984 and 2009

H. Gordon Green

Edited by Ronald Caplan

Breton Books

The first edition of *The Silver Dart* was published in celebration of the 50th Anniversary under the authority of the National Coordinating Council for the Golden Anniversary of Flight in Canada. Author H. Gordon Green dedicated that edition to Vernon Pope, "a gentleman and scholar whose extensive knowledge of aeronautics was of invaluable help to me in preparing the more technical parts of this book."

Editor: Ronald Caplan
Production Assistance: Bonnie Thompson, Sharon Hope Irwin
Layout: Fader Communications

Front Cover: Photograph by Marian Bell Whitcomb
Back Cover, bottom: 2009 *Silver Dart* by Vaughan Merchant
For their help with the photographs, our thanks to Valerie Mason, Doug Jermyn, Gordon MacRae and Eleanor Anderson. Other photo credits begin on page 136.

We acknowledge the support of the Canada Council for the Arts for our publishing program.

Canada Council for the Arts — Conseil des Arts du Canada

We also acknowledge support from Cultural Affairs, Nova Scotia Department of Communities, Culture & Heritage.

NOVA SCOTIA Communities, Culture and Heritage

We acknowledge the financial support of the Government of Canada through the Canada Book Fund for our publishing activities.

Canada

Library and Archives Canada Cataloguing in Publication

Green, H. Gordon (Henry Gordon), 1912-1991, author

The Silver Dart : the story of J.A.D. McCurdy, Canada's first pilot / H. Gordon Green. -- New edited edition.

ISBN 978-1-926908-29-8 (pbk.)

1. McCurdy, J. A. D. (John Alexander Douglas), 1886-1961. 2. Air pilots--Canada--Biography. 3. Aerial Experiment Association. 4. Aeronautics--Canada--History. I. Title.

TL540.M23G7 2013 629.13092 C2013-907254-3

Printed in Canada

FSC www.fsc.org MIX Paper from responsible sources FSC® C103567

Contents

Melville Bell Grosvenor
hanging from a kite rope

Introduction

I met John A.D. McCurdy in 1897 when I was visiting Baddeck and the Alexander Graham Bell home for the first time. A group of us were about to set out on one of the picnic excursions which were so delightfully frequent at Beinn Bhreagh in those days, and just as we came down to the wharf to load into our boat, I noticed a slightly built and scantily clad boy of about eleven standing in a canoe and watching our approach with evident amusement.

"How about a race across the bay?" he asked. "Here's another canoe." The challenge added to the merriment of the hour, but as for myself, I was somewhat too interested in the young lady at my side to accept. That young lady, as you may have guessed, was Elsie Bell, who later became my wife. In our party, however, were two young Harvard students, and one of them was foolish enough to get into the extra canoe which the boy was holding at the ready. He was beaten in the race by a most embarrassing margin.

It is with genuine pride that I can say that I have known McCurdy ever since that time and that I have been privileged to watch him closely as he has gone on through a most remarkable life of challenging and of victory. I knew him in the days of his famous houseboat when, with his brothers George and Lucian, they were teen-age buccaneers and sailed the Bras d'Or Lakes from one end to the other in their sloop. I watched him in the early days of kiting on the hills about that beautiful lake as he made use of his remarkable knowledge of sailing to help unlock the mysteries of flight. I followed him during the breathlessly wonderful days when the *Silver Dart* was lifting from the ice beyond Baddeck, and in all of the other marvellous exploits which this book records.

And I want to say how pleased I am that this book has finally been written and that future generations will now be able to read of the amazing achievements of this brilliant Canadian. I know that he is much too modest to tell that colourful story himself.

You know what my work has been. In my years with the National Geographic Society I have come to know nearly all of the great explorers of our time. I have known Peary, Shackleton, Amundsen, Lindbergh, Byrd and many others of their breed. And I want to say that I regard John A.D. McCurdy as a man who ranks with the very greatest of these.

When you consider that those early flights were made in the flimsiest of machines powered by the most primitive of motors, that there were no teachers nor books to serve as a guide, you will agree that I am not exaggerating in any way when I say that every time McCurdy lifted himself into the air, he took his life in his hands. And you will surely share my belief that he was as dauntless as Peary, or Lindbergh, or any of the other great explorers I have just named.

Nor were his achievements of any lesser consequence. It is my belief that few Americans are fully aware of how many and how tremendously important those achievements were to the progress of this century. I hope this biography will correct that lack and give McCurdy an honour that is long overdue.

This is a book which should have been written long ago, and I wish it well.

Gilbert H. Grosvenor, Chairman,
The National Geographic Society
Washington, D.C.
January 8, 1959

1

The Coming of the Great Man

It was 1885, summertime, and beyond the rim of the village of Baddeck, the waters of Bras d'Or were in a playful mood. In the stuffy office of the *Cape Breton Island Reporter* that day, the editor considered the lake's rollicking invitation and fretted because he had to struggle a little longer with his newspaper. He had been in a thorny humour all morning, for nothing he had done to make haste had gone right, and he was particularly irked by the fact that for no good reason at all, his telephone had gone into a silent sulk.

Now there was more than the irritation of inconvenience in the telephone's refusal to perform its customary miracle. There was a matter of pride as well, for in that year of 1885 the telephone was but nine years old; and the line which connected the newspaper office to the store which the editor's father kept, on top of the hill, was still regarded by the stolid villagers as an extravagant toy.

To the editor, the telephone was far too sacred a creation to have its innards probed by the hand of common man, and he was tapping it with a sort of fearful reverence when he noticed that he was being watched from the other side of the window.

The man who was curiously watching proceedings was someone Baddeck had never seen before. He was a well-dressed, well-built man, over six feet tall, with jet black hair and the good face of a man who is pleasant in spite of his obvious superiority.

Without waiting for an invitation, the traveller entered the printing shop. "Your phone won't talk back to you ?" he asked.

The editor explained that it had worked very well indeed till this morning, and for the life of him, he couldn't imagine what had got into it. "And the worst part is that there's probably not a man this side of Halifax that can fix it either," he added.

The big man smiled, went over to the telephone, tried it briefly, and then with a familiarity which made the editor flinch, he unscrewed the mouthpiece. He flicked a dead fly from the naked diaphragm, and put the phone together again.

"It will work now," he said.

Which it did, of course, because the visitor was none other than Alexander Graham Bell. And the editor he astonished that morning was Arthur W. McCurdy, the father of the man whose story this is.

So one can say with some measure of truth that it was a dead fly which first brought the Bells and the McCurdys together. Actually the dead fly merely brought the famous inventor over the McCurdy doorstep. But what brought Bell to Baddeck in the first place was a travel book—perhaps one of the first ever published about Canada—entitled *Baddeck and That Sort of Thing*. This was written in 1874 by Charles Dudley Warner and McCurdy always treasured a copy of it. In the main, the folder claimed three wonderful attractions for Baddeck: that it was beautiful, that it was Scottish, and that it was cool.

Bell, at 38 already one of the wealthiest men in all America, was searching for a place to build a summer home and he was intrigued. He had an artist's love of beauty, he cherished the traditions of his Scottish ancestry, and he had always abhorred hot weather. So he was exploring Baddeck now with his wife and two daughters to test the claims of the travel folder.

Arthur McCurdy was something more than an editor. He was a one-man Board of Trade, and he was quick to sense what a boon it might be to the community if the Bells could be induced to settle with them. So he hustled off a secret message to Mrs. Dunlop, who ran Baddeck's oldest inn, alerting her to the fact that the famous Bells were in town and would doubtless need lodging.

Would she be sure to have her best rooms ready and something special in the kitchen?

When the distinguished visitors eventually reached Telegraph House, they were pleasantly surprised to find rooms so spotlessly clean and so well aired. They stayed only three weeks that first summer, but that short time was long enough to fix a friendship between the two families which was to last a lifetime. It was the game of chess which continued what the fly and the folder had begun. Bell dearly loved to play chess, and Arthur McCurdy was the only man in the village who could play it. Better still, he was a good match for the inventor.

By the time the three weeks were up, Bell had completed his plans to return for a longer summer the next year. From grandfather McCurdy—owner of the store on the hill—Bell purchased a small farmhouse a mile and a half out of town, and left plans and orders for its remodelling and for the installation of a mysterious innovation called a bathroom.

The principle for the latter seems innocent enough to us now, for it was simply a device for collecting rainwater on the roof, storing it in an attic tank, and drawing from it at will to make a little whirlpool in a curious article called a flush toilet. The islanders, who came miles to see it had, however, another name for it. To them it was "the jinx," and they were not at all convinced of its propriety. The idea of performing one's animal functions in the very centre of so fine a house seemed to them to be neither sane nor sanitary. Besides that, they fancied the affair much too draughty.

Sane or not, the renovation of the old McCurdy farmhouse was completed on time, and the next spring, while all the village gaped, the Bells arrived from Washington with sufficient equipment to stay for an indefinite period. Included in that equipment was a phaeton complete with a spanking team of horses and a coloured coachman named Perrin. To Baddeck, the jovial coloured man was even more intriguing than the jinx had been.

From his very first days as a resident, Bell fitted in well with

Baddeck's community life. Sudden fame and wealth had never destroyed his love for fellowship, and there was a persistently human quality about the man which enabled him to enjoy the companionship of a humble farmer quite as well as that of some of the great scientists who often came from Europe and the States to visit him in his new retreat.

But though he became friend and fairy godfather to the whole village, it was the McCurdys who continued to be closest to him, and that same year Bell induced Arthur McCurdy to leave his paper and become his secretary. It was an admirable relationship, and one which withstood the tests of sixteen years.

In that year of 1886 the most memorable event so far as this book is concerned was the christening of Arthur McCurdy's second son, John Alexander Douglas, in the Presbyterian church. Though he was no churchman, Bell attended the rite, and the cup which he had engraved for the occasion is now one of the most cherished of the McCurdys' heirlooms.

And indeed it should be cherished, because it marked the beginning of one of the most fascinating and most fruitful associations in all Canadian history.

The young McCurdy—he went by the name of Douglas at home—knew of those days when the Bells were at the farmhouse only from hearsay. By the time he was old enough to see those things he was to remember, Bell had moved to the beautiful 1600-acre estate across the bay to which he had given the Gaelic name "Beinn Bhreagh" (beautiful mountain). Here he built the palatial mansion which still stands, and his laboratories. It was here too that, to power those laboratories, he installed Cape Breton's first dynamo, feeding it with the flow from a spring discovered higher up the hill.

"I guess the earliest thing I can remember about the days when I was a kid roaming around the Bell place," McCurdy said, "is the way the brass turnings used to spiral off the lathe in the lab. Those brass shavings were wonderful treasures."

He also remembered the mornings when he and Bell's daughters would go upstairs to rouse the great man from his four-poster bed. Bell had the habit of rising late and since it seemed that he was always in a mood to play, children frequently pestered him. McCurdy remembered well how Bell would draw his knees up under the blanket, allow the children to take turns "climbing the mountain" and then suddenly open his knees to bring the climber down from the heights.

He remembered, too, the Sunday night songfests which were a tradition at the Bell mansion, with Bell himself at the piano. It was a tradition which McCurdy was later to bring into his home.

And he recalled with a smile, how the Bells dressed whenever they gave a party at the big house, with Mrs. Bell in flowing Grecian-type garments, and Bell himself in knickerbockers, red tie and velvet jacket.

But best of all, he liked to recall the Wednesday evenings when the Bell house was the scene of something like a thinkers' club, and the neighbours were all welcome to attend. These unique meetings had become a part of Bell's way of life long before he had discovered Baddeck, and when he held them at his home in Washington, some of the great minds of the day often took part in the discussions.

The idea now came to Baddeck, and every Wednesday which found Bell in his summer home was adult education night. First there would be checkers or whist, then sandwiches, ice cream and drinks in the dining room. Since Bell was a teetotaller, the drink was nearly always raspberry vinegar, of which Bell himself was very fond.

Then after all had eaten, the group went back to the living room to discuss whatever topic Bell had chosen for the evening. And since the Bell home was rarely without some distinguished visitor, some of the themes were given the benefit of eminent authority. None the less, the shyest villager present was asked for his opinion too.

It was an institution which made a profound impression on the young McCurdy, and one which he followed with intense interest all of his years in Baddeck. Bell kept up the practice as long as he lived.

There was often some applied science at these gatherings, too, and McCurdy liked to think of the winter when Bell was absorbed with his experiments in recording, and used his Wednesday night guests to help him.

It is not generally known that it was his cousin Chichester who, along with Bell, invented the wax record, an idea which was later sold to Edison. (Edison was still trying to perfect tin foil surfaces.) But the wax record was an idea which probably originated right in Baddeck. Certainly many of the first records were made there, for Bell got some of the villagers to sing in Gaelic and then astounded them by playing the record back to them.

Bell had a keen sense of humour, a quality which Baddeck appreciated to the full, but there were times when it was somewhat difficult for the unenlightened to decide where the great man's humour ended and where his science began. There was the time, for instance, when at the beginning of his dinner he suddenly called for a bucket. A glass jar was found instead and Bell placed it on the table beside him. Then into the jar, as methodically as if he were in the laboratory, he began spooning the various available foods which were being served. He allotted them in the same proportions that he thought he would be likely to give to his own plate, added the proper kinds and amounts of drink and then stirred the whole mess together.

"I just wanted to see what a man's stomach must look like after a meal like this." He said it cheerfully enough and the cook didn't know whether she should laugh or cry.

But perhaps the drollest of all the stories McCurdy liked to tell about his boyhood days at the Bell house is the one about the great cat experiment.

In the summer of 1900 there were two famous scientists visiting Beinn Bhreagh. From the Smithsonian Institution in Washington came Professor S.P. Langley, who was at that time weathering the ridicule of almost every other great scientist in the land because he believed it entirely possible for man to achieve flight in a heavier-than-air craft and he kept reiterating that belief with an almost plaintive insistence.

Langley had, in fact, already flown a model of the craft he proposed to build to carry a man, and Bell had witnessed that flight and had been profoundly impressed. But the successful flight of that model seemed to have answered none of the critics, and one of the most learned of those critics was Nova Scotia-born Professor Simon Newcomb, perhaps the greatest mathematician and astronomer of his day and then a professor at Johns Hopkins University.

Professor Newcomb had his own carefully reasoned answer to Langley's optimism about the possibilities of flight, and he had worked out the theory known then as the "Law of Squares and Cubes." In effect this law pointed out that when the dimensions of a wing surface were increased, for every instance in which the surface was squared, the weight would be cubed.

The truth of that law, asserted Newcomb, gave an explanation as to why, though Langley's model might successfully achieve flight, the larger craft he proposed was doomed to failure. It would have to be far too heavy.

Bell was well aware of the sharp enmity between the two men and while he had invited both to visit him, he hadn't the slightest intention of letting them come at the same time. It happened, however, that Langley arrived in Baddeck some considerable time before he was expected, and Bell discovered the fact too late to avoid having the two men meet. But for all of his genial efforts to keep the meeting free of bitterness, it was evident right from the start that a clash was inevitable. Langley was a rather quiet and extremely sensitive sort, but Newcomb was a man who could not be easily contained. He also undoubtedly thought himself the more famous of the two.

Now it happened that Beinn Bhreagh abounded in cats. Melville Bell, father of the inventor, was a great cat lover, and he had a score or more of them to love whenever he came to stay with his son. So Bell, in his rather nervous efforts to find conversation topics which would not lead to an argument between his two guests, was making some lighthearted jest about the disgracefully high birth rate of the cats on his estate, when Langley suddenly said, "You know, it's a strange thing why a cat always lands on its feet when it falls."

Newcomb was upon poor Langley in an instant. "What do you mean by that?" he demanded.

"I mean that if you take a cat, turn him upside down and drop him, he invariably lands on his feet. Nor does it matter from what height you drop him."

"Impossible!" Newcomb declared. "Absolutely impossible! And any scientist should know that it's impossible!"

"Why should it be impossible?"

Newcomb explained the impossibility with cold science and heated indignation. In order for anything to be turned, there had to be a fulcrum, and where could a cat in mid-air find a fulcrum?

Bell, always a peacemaker, saw a storm about to break and took prompt steps to avoid it. Quickly he looked around for old Duncan B'hanan, the houseman. "Duncan!" he called. "Bring us a clothes basket!"

A little bewildered, Duncan went into the laundry room and came out with a large wicker clothes basket.

"Now then, collect us a basketful of cats. Collect every cat you can find on the place!"

His bewilderment increasing, Duncan began calling the cats and putting them into the basket. When the cats began to protest the indignity with too much vigour, the three scientists all took a hand in preventing them from escaping. And somehow the four men and a dozen cats reached the balcony which overlooks Bras d'Or from the west side of the house. That balcony is over twenty feet high, and when Bell took a second look at the extent of the drop, his kindliness made him hold up a delaying hand.

"Duncan!" he said. "Go downstairs and collect up all the cushions you can find and lay them on the terrace. And whatever you do, mind that Mrs. Bell doesn't catch you at it!"

Mrs. Bell was not only the business head of the house, she was its disciplinarian as well. And Bell knew that better than anyone.

So eventually the outraged cats were taken from the basket and dropped one by one to the cushions below. And the three world-famous scientists, their bearded faces grave with the ponderous import of the problem, watched and compared observations.

Langley was right. No matter what position a cat was in when he left the hands of the great men on the balcony, it was his feet which first found the comfort of Mrs. Bell's cushions.

Newcomb was a big enough man to apologize afterward, and thanked Langley for giving him a new problem in physics to resolve. A cat did turn in the air without needing a fulcrum. But how?

Langley, however, couldn't be bothered with anything so piddling. He was much more interested in solving the problem of how to break the bounds of earth and lift himself into the air with the birds. And there seems as much irony as wonder in the fact that a bright-eyed lad of fourteen who, from a discreet distance, watched the scholarly dropping of the cats that night, was to achieve the trembling freedom of birds before he would.

That boy was the curiously mischievous one of the McCurdy lads whom Baddeck called Douglas.

2

The Houseboat

The life of the young McCurdy was so closely linked with that of the Bells that there has always been a tendency on the part of many Canadians to think that his fame was something which Bell willed on him, rather than something which is rightfully his own. Nothing can be farther from the truth, as a careful study of McCurdy's eventful years will readily prove. It is true, nevertheless, that Bell's interest in the young lad was touchingly paternal. When the boy was five, the Bells wanted very much to adopt him, and had it not been for his strong-minded and motherly maiden aunt, Georgina McCurdy, Douglas would have undoubtedly become the Bells' legal son.

The reasons for this are easy to understand. Douglas could never remember his mother, for she died before he was two, giving birth to his brother Lucian. The Bells, on the other hand, had lost their only two sons in infancy, their two daughters were fast growing into womanhood, and there seemed little chance of any newcomers to take their place. Bell delighted in having children about him and he had always had a special liking for the boy who seemed forever watching the unpredictable variety of experiments which went on in the Baddeck laboratories. Furthermore, with Arthur McCurdy as his secretary and constant companion now, the boy would never have to be far away from his real father.

In 1893 the Bells and the McCurdys agreed upon a sort of trial adoption, and when the Bells packed up for the fall trip back to their Washington home, and Arthur McCurdy prepared to go with them, Douglas was scrubbed and togged for the trip too.

Of that winter's stay as a member of the Bell family in Washington, McCurdy had pleasant enough memories. He went to a private school that winter, snowballed the kids down the street, enjoyed the companionship of the Bell family as well as that of his father and, as usual, he was often watching from a quiet corner somewhere when greatness called to consult with Bell. But for all the pressure which Bell's fame and position put upon him in Washington, he found time for the lad he hoped to make his own.

One of McCurdy's clearest recollections of that Washington year was when Bell took him to watch the turning of the first sod of the Volta Bureau Institution which the inventor had endowed with money received from the sale of his wax record patents to Edison. It was founded in 1887 by Bell for the increase and diffusion of knowledge relating to the deaf.

But if the young McCurdy was indifferent as to whether he was to belong to the Bells or not, his Aunt Georgina was not. The winter had apparently been a lonely one for her, despite the fact that she had saddled herself with the care of the baby Lucian, Douglas's older brother George, and a sister, Susan. But she wanted Douglas too.

It wasn't right, she thought, to break up the family, and now that she was in the business of raising a family anyhow, what difference would one more child make?

There was something else in the back of this shrewd woman's mind which she couldn't conveniently explain to the good man who wanted Douglas for his own. She had been reared in the old Scottish belief that the devil soon finds work for idle hands to do. She believed that for a boy to make the acquaintance of the dignity and discipline of work was just as important to him as his school lessons, or his catechism, or his porridge. And what work would Alexander Graham Bell have for a boy in a big, rich city where they didn't even have woodboxes?

So Douglas came back to Baddeck to live with his aunt, his grandfather, his sister and two brothers. For the summer, he would have his father, too, and what more could he ask for in the way of companionship?

But like any other lad who belonged to Cape Breton in those days, he did find fault with the work which had to be done. Besides the little store which Grandfather McCurdy ran, he had a farm with a woodlot, two cows, a cart horse and three Sable Island ponies. The ponies were there mainly for the sake of the boys and perhaps to provide more reason than usual for requiring the boys to do the chores.

School in those days was from nine to four, with but a single hour off at noon. In that hour the McCurdy boys had to hustle home, eat their own dinner and then feed and water the stock. They were also required to feed it before breakfast and after supper, to milk the two cows, and bed everything down for the night.

That round of duties did not excuse them from the requirements of the house, however, and it was their bounden duty to keep their aunt amply supplied with coal for the kitchen stove and wood for her furnace.

There was no variation in this rigid schedule, no excuses, and the years only added to the chores required. By the time Douglas was ten, he and his brothers were cutting wood in the bush as well, and putting up the hay.

But by the Cape Breton standards of the time, Grandfather McCurdy was probably not so strict as he might have been. He was the chief elder of the Presbyterian church, held morning and evening prayers for the family, demanded that each should know his shorter catechism thoroughly, and always said grace. Nor did anyone leave the table before he had finished his meal.

Still he allowed the boys a freedom which must have seemed surprising to his neighbours, and perhaps even dangerous. Old-timers around Baddeck recalled the houseboat, the strange craft with which the McCurdy boys braved the challenge of Lake Bras d'Or when the oldest of the trio was only twelve and Douglas but ten.

This boat was quite frankly inspired by an equally strange

craft which Alexander Graham Bell one day introduced to Bras d'Or. It was a boat of the catamaran type which was towed in a leisurely fashion about the lake when the summer season opened. She had upper and lower berths, a veranda, an upper deck and most wonderful of wonders, another "jinx." The McCurdy boys were often asked to accompany the Bells when that family picnicked and fished on this craft. And the lads were so tremendously impressed with the Bell houseboat that they made a solemn pact to make their own.

The boys' craft turned out to be a Huckleberry Finn type of raft with a platform floated by eight kerosene barrels. Pitched on the platform was a hut with a canvas fly built above it.

Regardless of what the neighbours thought of its rightness, the boys considered it the high point of their careers and lived and slept in it all summer long. And Arthur McCurdy must have been impressed too, because he decided that each of his three sons should keep a diary of his adventures on the high seas. He offered prizes for these. First prize was $3.00, second was $2.00, third was $1.00. George, the oldest brother, always won. Douglas was always second, poor Lucian was always honourable mention. The diary kept by Douglas McCurdy in the summer of 1898 is reproduced in full starting on page 123.

Mrs. David Fairchild, the younger of Bell's two daughters, has her own special recollection of that famous houseboat. Once, during that summer of its launching, she spent a night aboard. Though there was no chaperone, the rules of propriety were carefully observed. The single bed, which the boys had designed for themselves, was given to the guest. The three sailors slept on the floor outside and the tablecloth was hung up between them as a partition.

In the morning, Mrs. Fairchild recalls that from the boys' side of the tablecloth, George gently advised her, "Don't get up yet, Daisy. We'll tell you when."

Then she heard three splashes echo through the morning air as the McCurdy boys took the plunge which always christened the day for them. They clambered aboard again, there were a few muffled sounds of battling over the towel and socks, and then once again, George's gentlemanly voice. "It's all right now, Daisy. You can get up."

But there was one inescapable day of reckoning to each of these poetic summers. When it came time for haying, the houseboat had to be tied to a stump and the three crestfallen buccaneers had to go up to their grandfather's fields to cut, rake, pitch, and sweat till the last precious forkful was tight against the rafters.

Years later McCurdy said, "I can still feel the thistles between my shoulder blades!"

3

A Lesson in Caution

But while most of Douglas McCurdy's boyhood life was idyllic, carefree and healthy, there was one momentous incident in it whose terror was such that he would flinch to recall it. The scene of the tragedy was the big frame house on the hill where Grandfather McCurdy and Aunt Georgie were doing their painful best to raise the orphaned family in the way it should go. But the idea which sparked the disaster was born in the Bethune house, a catapult's shot down the road. There, in a fine old house which still stands, Doctor John and Mary Bethune were also experiencing the gentle agonies of raising lively boys. Their oldest son Robert was exactly nine that day, Rodderick about seven. And the pair were constant companions of the three McCurdy boys.

It was a Saturday in October, the year after Douglas had come back from his Washington winter with the Bells. The lake was too cold to be enjoyed, and life in Baddeck was just a trifle less active than usual that day. So when Mrs. Bethune suddenly announced that she would give a birthday party for the five boys that night, it was genuine cause for celebration.

Mrs. Bethune was a very well-organized soul and had planned the affair quite completely. There would first be a game of hide-and-seek, then a round of football with a brand new pig bladder, and finally, when it was time to pay attention to the appetites such exercise would generate, there would be a beribboned table loaded with cookies, cocoa and birthday cake.

"How about some fireworks ?" The eager question came from Douglas.

A swell idea indeed! But where in the world would you find fireworks in Baddeck ?

But Douglas had thought of the answer before he had posed the question. Up in the attic there was a little room where Arthur McCurdy, the boys' father, kept his treasured collection of guns and hunting equipment. It was a room which was strictly out of bounds to anyone but himself, for Father McCurdy had no illusions about the curiosity of his sons and he had to be away from home a great deal of the time. Furthermore he had always been a great hunter. Indeed it was he who shot the last three caribou ever reported on the Island, and the fine head from one of those animals still looks down from the wall over the great fireplace in Beinn Bhreagh. So the door to his gun room was sealed with a lock which, so Father McCurdy thought, would make the room impregnable.

But even at the age of eight, nothing ever seemed impossible to the young Douglas McCurdy. He engineered the picking of the lock, just as he had engineered all the other parts of the idea. The boys had a fine time in the room before they got down to the more exacting business of manufacturing fireworks.

"All we need is gunpowder," Douglas said, "and there's lots of it here. There's some in every shell."

So it was that the boys gathered up some high-calibre bullets, sneaked them downstairs and out into Grandfather McCurdy's workshop and there they put them in the vice and worked the lead out of them. Then the gunpowder was poured out on a newspaper and taken over to the yard behind the Bethune house. Here, Douglas dipped out a bit of the precious powder, put it in a tiny heap on the ground and gingerly touched a match to it. The result was a sudden "poof" of flame and smoke which left a delectable odour hanging in the air.

Then he tried another charge and another. Next a sort of gunpowder snake. The effect was tremendously satisfactory to all but Douglas. For him it was not enough.

"For a party like this, we want something that will make a big bang!" he said.

The solution was simple enough to Douglas. They would have to use more powder and they couldn't flare it in the open. It would have to be least partially confined.

Back in the forbidden attic room, hanging on the wall along with other curios which were almost as interesting, was Father McCurdy's powder horn, a relic of muzzle-loading days. Douglas went back for the horn, inserted the plug into the tip and took it to the Bethunes' yard. With a calculating eye, he gave it enough powder for what he thought would be a "big bang," took the horn in his hand, held it as far away from him as possible and dropped a lighted match into it.

The roar was one which rattled the windows across the street, and when the smoke cleared, his terrified pals found Douglas in the Bethune ditch as motionless as death itself.

When Aunt Georgie and the first of the startled neighbours began to arrive, they found Douglas burned black all down his right side. His eyebrows were completely gone, his jacket had been blown to tatters, the right side of his face could scarcely be recognized, and the hand which had held the horn was little more than raw flesh.

With Aunt Georgie and the village doctor in charge, Douglas was carried back to Grandpa McCurdy's, and there, in the very room where he had been born just eight years before, the family waited for him to die. But he was still alive next morning, and there were special prayers in the Presbyterian church offered for him in both the Sunday School and the regular service.

And when the first few terrible hours had passed and it became evident that the boy's vigour might possibly pull him through the ordeal, there was still the fear that he would lose the sight of at least one eye. The room was darkened and the eyes were kept carefully bandaged for several days. That fear was a short-lived one, but a greater one took its place.

While Douglas's jacket had given some protection from the blast to his arm and shoulder, his right hand had taken the full force of it, and was horribly damaged. And late that first week, the doctor discovered unmistakable signs of gangrene. Now thoroughly alarmed, he called in two other doctors. A decision to amputate in one so young is always a terrible one for a doctor to have to make. And in those days when the technique of modern surgery was still a long way off, it was a very desperate step indeed. But the doctors agreed that there was nothing else to be done, and sorrowfully they gave the news to the family.

"How much can you leave?" Aunt Georgie wanted to know.

"We will have to amputate at the wrist," they said. "And we will have to do it soon."

What they did not discuss with the elder McCurdys was that with Douglas already in a shattered condition, there was barely an even chance that he would survive the shock of such an ordeal. But they did not have to tell Aunt Georgie. She was an excellent nurse and a seasoned one, and her estimate of the chances for success was probably as shrewd as those of the doctors.

"When do you want to do it?" she asked. "And where?"

The doctors gathered heads again.

"Tomorrow at eleven and right here," they said. "We don't dare move him."

So, her face firmly set against any further weeping, Aunt Georgie got the room ready for the operation. The news spread like a trail of windblown fire through the countryside and next morning the village schoolmaster set aside the morning's classes and paraded his pupils in a body up to the McCurdy house on the hill. There, a file was formed and one by one the whispering, awe-struck children were ushered into the bedroom to see their playmate before he went under the chloroform and the knife. The schoolmaster had warned them that they must only wish Douglas well, but even those who were very young had the feeling that the real reason for this hushed visit was to say good-bye.

With the school visit over, the doctors bustled about with their last-minute preparations, checked and rechecked the supply of boiling water Aunt Georgie had for them in the kitchen below, and the smell of disinfectant began to fill the room like a faint fog.

But about a half hour or so before the appointed hour, the sound of a carriage and fast horses rang along the driveway outside and stopped with a tense "whoa" in the yard beneath. A moment later, his voice thundering like his giant footsteps, Alexander Graham Bell strode into the room.

"I want to talk to you about this operation," he said to the doctors. "It's been on my mind all night long. Are you *sure* you have to do it?"

Now, from the day when he had first come among them, Bell had always been considered by the people of Baddeck as something like a village godfather. So that his views and advice would have found a hearing at any time and with any of them. But Cape Breton's medical men had a special respect for Bell because, technically speaking at least, he was one of them. An interesting fact about Bell which our present day historians seem to have forgotten was that in 1886 he was granted an honorary degree of Doctor of Medicine by the University of Heidelberg at its 500th anniversary. The degree was probably the only one of its kind ever to be granted, and was conferred for his invention of a telephone probe for the painless detection of bullets in the human body. So Bell was really a medical doctor and as such he had the right to talk to other doctors, and on the same level.

The inventor was surely one of the most versatile scientists the world had ever seen, so it is not strange that he should have been a great believer in the ability of nature to be her own physician.

"Now, is this hand any worse today than it was yesterday?" he quietly asked his confreres.

The doctors were not quite agreed on that point.

"Do you think the boy's life would be endangered if we postponed this operation one more day?"

The doctors looked at one another and finally agreed that a day's postponement would probably not be dangerous.

"Well then, let's have a look at this hand tomorrow and see if there is any change. If there is, and the change is for the worse, it will be time then to talk of taking the hand away."

The advice of the great man was considered carefully and the doctors finally nodded assent. "We will wait for another day then," they decided.

There was no change for the worse next day, nor the day after. The doctors prescribed dressings of lard and boracic acid, and Aunt Georgie changed the medications every day, blessing each with her prayer and her tears. The recovery was a slow and a painful one, but it was complete. Six months later, Douglas McCurdy had a brand new set of eyebrows, his face was as fresh as ever, and he had regained the full use of his right hand. But the most important part of his recovery was one which the doctors probably did not notice. For the boy's curiosity and his love of adventure were still unshaken. Only now he had acquired the proper caution to go with them.

At the time of the celebrations of the 50th Anniversary of his flight in the *Silver Dart*, McCurdy was 73 years old—the world's oldest living pilot. All of the other intrepid adventurers who rode their crude birds into the untracked skies with him had long since passed from the sight of men. Was it luck alone that had enabled McCurdy to survive them? In modesty he may have said that it was luck, but perhaps the caution he learned in the Bethunes' back yard that October Saturday had something to do with his success too.

4

Baddeck and That Sort of Thing

One cannot read very far into the young Douglas's diary without sensing that the Baddeck of 1900 was a pleasantly remarkable community. Nor was its unique quality entirely due to the leavening influence of the Bells. The sterling virtues of the Canadian Scot had already given Cape Breton a wonderfully reliant character all its own. And Douglas's Huck Finn prose of that boyhood summer on Lake Bras d'Or shows us that very clearly.

There were no community youth projects then, no scientific supervision of a boy's leisure, not even a Scoutmaster in town. But there was a happy atmosphere about that old-fashioned way of life which allowed a lad to do pretty much as he wanted so long as he got in no one else's way.

Many of today's eagerly conscientious parents—that kind at least who raise one neurotic brat and then consider themselves experts in child training—will no doubt think that such a disregard of a twelve-year-old boy's activities is little short of criminal. Surely it would be difficult for them to believe that any self-respecting parent could sit unconcernedly at home knowing that his boys were out on the lake in a gale, and on so clumsy a craft as the houseboat was. But the senior McCurdys were not unconcerned. Their frequent visits to the houseboat constitute good proof of that, and so does the fact that Arthur McCurdy insisted upon each boy writing a diary of his adventures.

Baddeckers simply had the calm good sense to act as if their children were almost as smart as their parents, and almost as capable of looking after themselves. And what else could a youngster do with such a belief governing him but to try to live up to it ?

Certainly the lack of restrictions on the McCurdy boys gave them more than the usual chance to become creative. Yet they had also been given a very healthy introduction to hard work. And they had grown up in homes where the ideas of responsibility and frugality were little less than a religion. So it is little wonder that they could build so well and with so little help from the folks at home. Nor is it strange that they should have become remarkably self-sufficient and so lacking in fear.

To stumble upon youth grown out of such a frame was surely as fortunate a find for the great man from Brantford and Washington as was the tetrahedral cell or the dynamo. And it was of far more importance than the fact that he had at the same time found that particular kind of geographic beauty for which he had looked so long.

"I can never recall being afraid in the early emergencies of flying," McCurdy said later, "I was too busy thinking of what I had to do."

Fresh from a society which was already getting brittle with intellectual pretension, Bell must have been intrigued also with the unusual insight of the Cape Bretoners and the way that they read. As Douglas' diary shows, Cape Bretoners read often and they read deeply.

"They are not likely to have so many books on their shelves as we do," a Montreal critic once said, "but what they have will be good. And when they have finished with the book, they will have a knowledge of it that would put a professor on the defence."

So it should not come as a total surprise for Canadians to learn that Baddeck had claims to greatness besides those of being the location of Bell's countless experiments and a hatching place of great flyers.

A social club was founded in 1891. The original idea of starting such a club in Baddeck came from Mrs. Bell. In Washington a little earlier that year, she had been a guest at the home of Senator and

Mrs. Hale of Maine. According to the Senator's political critics, Mrs. Hale had all the brains the family possessed. In any case she was a woman with far more intelligence than most of her sex dared display in that gentle age, and she was in the habit of inviting kindred souls to a weekly discussion on matters of the day. This was by no means a mere feminine gossip group. The topics which the women tackled were of an importance traditionally thought to require the analytical minds of men. Mrs. Bell was so impressed with the idea that she immediately wanted to transplant it to Baddeck.

She could not have wished for a more willing soil, but the concept was of such novelty and dimension that the Baddeck ladies were quite overwhelmed to know how to put its high aim into suitable words.

As Mrs. Bell wrote in one of her papers, "such discussions we had over that—what serious seances were held in the big living room in the Lodge where we then lived, in the row boats, and even in that sacrosanct building—the laboratory. Finally we thrashed out the general idea, but how to express it in a few terse words? Over that we struggled again, and it was only after many days and hours of anxious deliberation that at last Dr. Bell wrote out the declaration for us. 'Its object shall be to stimulate the acquisition of general knowledge and to promote sociability among the young people of Baddeck.'"

Drawing up the articles of a constitution did not take quite so much mental agony. Bell cautioned the good ladies about the dangers of imposing fines for non-attendance.

"No one should come to a study group just to escape paying a fine," he said.

Bell also was the instigator of the wisest of all the bylaws—that which forbade more than five minutes discussion after the reading of any paper.

Current events were the most popular topics, with the Bell house invariably supplying maps for the occasion. To this day, the club has the curious tradition of always having the speaker read her paper while sitting down. This was done originally at Dr. Bell's request so that his wife, whose deafness would not permit her to follow the paper otherwise, could look over the reader's shoulder and follow the paper sheet by sheet as it was read.

As long as she lived, Mrs. Bell was the honorary president. The first president was McCurdy's beloved Aunt Georgina. On the first board were Mrs. J.L. Bethune, Mrs. T.S. McLean and Mrs. H.M. MacKenzie—names still fondly remembered in Baddeck.

The first name of the group was "The Young Ladies Club of Baddeck," but a few years later the members took cognizance of the inroads of age and remembered their avowed intention of being forever honest, and so changed the name to the "Alexander Graham Bell Club."

It is under this name that the club still functions. It was the first club of its kind in the nation.

In the 1890s, visitors of memorable importance descended upon Baddeck, for the little village was privileged to entertain the Governor-General, Lord Aberdeen, and his wife. And to make the occasion completely glorious, the vice-regal couple were accompanied by the Prime Minister, Sir Wilfrid Laurier. Even at this rather early stage of his leadership, the colourful Laurier commanded a tremendous appeal and respect, and Cape Bretoners came from both ends of the island to see him. The young McCurdy did better than that. Somehow or other he was able to shake the great man's hand, an incident which seemed of such magnitude to him that he could hardly be moved to wash his hand that night.

But to Aunt Georgina, a greater honour was to come. When Lady Aberdeen expressed a wish to do some travelling about the picturesque island while her husband visited with Bell at Beinn Bhreagh, McCurdy's aunt, as President of the "Young Ladies Club," was chosen to be her escort. So in a carriage drawn by the best horses that could be borrowed, Aunt Georgie drove the Governor-General's wife around the country roads leading out of Baddeck.

A rather intimate friendship grew up between the two ladies on that trip, and there seemed no doubt that Lady Aberdeen's interest in the Cape Breton way of life was genuine. She liked to visit people as well as places, and was visibly moved when introduced to invalids and the hopelessly aged. She seemed particularly interested in knowing what happened to the isolated farmer when sickness or accident struck.

"You have so few doctors!" she said to Aunt Georgie. "And some of these people are so far away! What do they do when emergency strikes?"

Aunt Georgina had to admit sadly that an emergency or illness could be a very tragic thing in Cape Breton's less accessible corners. But then, one took that chance when he chose to live so far away from town.

For all its breath-taking beauty, Lady Aberdeen could not forget the island's desperate need of more medical care. To state simply that there should be more doctors was hardly the answer. Many of these people could not have afforded the expense of a qualified practitioner even if he had been available.

"But there should be some way!" the great lady thought. "Why couldn't we have nurses trained for taking care of cases in the home? Nurses who would be willing to travel like a doctor?"

As the friendship deepened between the two women, an idea began to take shape. That idea continued to grow long after the Governor-General's wife had returned to Ottawa—even after the Aberdeens had finished their term in Canada and were back home in Scotland.

Aunt Georgie made two trips to Scotland in the next six years to help work out the details of organization. The idea became the reality which is now known as the Victorian Order of Nurses.

Baddeck pioneered still another notable movement. The Parents Club came as a sort of project promoted by the original "Young Ladies Club," and here again Dr. Bell was called in to give his ideas as to what might best be accomplished by the formation of a parents' association. It was the first such organization in Canada.

"The parents bunch," McCurdy recalled whimsically, "didn't really get going strong enough in time to do me any harm."

5

The McCurdys Go to Toronto

In 1903, Baddeck's little frame school had given Douglas all the education it was capable of mustering, and though his pranks and restlessness had given his teachers many an anxious moment, none of them had ever accused him of being stupid. His principal, a hard-working, underpaid Scot by the name of James McPhee, was the most insistent of all.

"Aye, that lad must go on now," McPhee told Douglas' father and Aunt Georgie. "He is foredestined to go places, that one!"

So with all of his teachers, all of the McCurdys, and all of the Bells in complete agreement, plans were made for Douglas to go on to higher learning. There were two good enough reasons why Toronto was the university elected to be honoured with the young lad's presence. In the first place, Douglas was quite firm in telling the family that he wanted to be a mechanical engineer, and the University of Toronto had the proud habit of claiming that its engineering school was the best in the land.

Secondly, the McCurdys had relatives in Toronto, and Arthur McCurdy, who was a seasoned traveller himself by then, thought it time for his children to discover that Baddeck was not really the only cultured place in America. It was fitting, he thought, that they see a bit of the world now.

So it was that the young McCurdy, firmly armed with a Nova Scotia high school certificate, presented himself at the gates of Toronto's famous university. He was only sixteen at the time and the tenderness of his age made the solemn men at the desk look up at him with sympathetic concern. Furthermore, they were not at all sure that credentials from a little country school in Nova Scotia were really what their university required. But a careful questioning of the boy and a special examination brought him in with flying colours, the youngest student to be enrolled that year.

McCurdy does not recall those university years with any particular thrill, nor more than the usual scholastic pride. He was a better-than-average student, but disinterested at times, often impatient, but a pupil who always passed examinations. Of the lessons he sweated so hard to learn, he could recall but a few, and he sometimes wondered if even those remembered ones had ever done him any good.

But Toronto did give him some wonderful friends, and one of these was destined to share the adventure of his life so constantly that the world would link their names together for a generation or more. That friend was F.W. "Casey" Baldwin, a great all-round athlete and one of Toronto's most pampered football players. He was a grandson of Sir Robert Baldwin, Prime Minister of Canada prior to Confederation.

The friendship which grew up between McCurdy and Baldwin was one which not only withstood the test of time but the more difficult ordeal of fame as well. There was to be ample room for disagreements and jealousies in the years ahead, but nothing of that nature ever came between the two, even for a moment.

There was one other happening of those college years that McCurdy remembered with almost equal warmth. On a trip to Montreal with his football buddy for a McGill-Toronto game, McCurdy was astounded to see his old friend Mr. James McPhee in the McGill line-up. At long last, the hard-working Scot had finally been able to enroll in the college of his choice. So there was a gala reunion that night in Montreal. McPhee was a freshman in McGill, McCurdy in his third year at Toronto.

"So he wasn't Mr. McPhee any more," McCurdy recalls. "Now I could call him just plain Jimmy."

Jimmy McPhee was foredestined to go places too in the years

to come. He became so celebrated as a geologist that Cecil Rhodes induced him to come to South Africa, where he became a leading name in that country's mining world, and died there a famous man.

But if the Toronto years were to bring Douglas such valuable friendship, they were also to bring him a touch of sadness. Most of the family had fitted into the Toronto environment without too much effort, but with grandfather the change could not have been an easy one. He was ninety-three now, still gallant, still alert, and in full possession of all his faculties. But he had lived too long beside the beautiful shores of the Bras d'Or, and had known its people too well. And now that he felt the sunset so near to him, he was far more homesick than he wanted anyone to know.

But his work was not quite done, and in one of those secretly lonely hours in Toronto when he was trying hardest not to be homesick, the old man took a sheet of Douglas' examination foolscap, and in a hand heavy with age, but still quite firm, he wrote a letter which the McCurdy family regarded as one of its most treasured heirlooms.

"To My Four Grandchildren

"Toronto, March 8, 1903, Lucian's birthday

"To Susan, George, Douglas and Lucian, my Great Grand Children.

"I thought you might be pleased to know some incidents of my early life that you have not as yet known.

"Born in 1810, July 29, in Onslow, Nova Scotia, my grandfather was an immigrant from Londonderry, North of Ireland. When my father grew up to manhood he married Agnes Archibald. He purchased a large farm in Onslow where he spent his life as a farmer. They were prosperous and happy. Their children numbered fourteen, seven daughters and seven sons, myself the youngest son, remaining with them on the farm until I was 22 years old, at which date I married a beautiful girl and of course I was very happy for she proved herself to be *good, better and best.*

"Before being married a year I bought a farm and agreed to pay five hundred pounds for it with interest till paid. It was quite a venture but we agreed to face it and do as well as we could. We went vigorously and from year to year we paid more or less so that in the course of twenty-five years we paid the whole amount with interest. You may be sure we worked hard but we were always cheerful and happy. Your Grandma did her part nobly without the slightest complaint. In due time sons and daughters came who were a blessing to us as we toiled with considerable cost. They were educated in the old fashioned way but notwithstanding they were always respectable and kept good society making themselves useful.

"And now my Grandchildren what shall I say? How well I remember when your dear loving Mother was married, so nice she was, I always loved her. She lived in the house with us and gladly came in to our family worship when your father was away. Lucian was only a few days old when your Mother died, shortly before that she committed you all to Aunt Georgie's care by whom you have been carefully trained up to this day. Susan may perhaps remember some of her Mother's last words among which 'Susan, I am going to Heaven, be sure and come there also.' No doubt she wished the same for you all. You cannot be thankful enough for Aunt Georgie's care and kindness. God will reward her and honour her, whether you do or not.

"God bless you my dear children and prepare you to go to meet your Mother in Heaven where she has been for so long a time so happy.

"Your Grand Pa"

6

Aerial Experiment Association Ltd., Baddeck, Nova Scotia

Despite the distance between them now, the McCurdys and the Bells did not allow themselves to grow apart. And in one of her many many letters to the eager young engineering student at Toronto, Mrs. Bell said: "And now Douglas, you will be coming back to us soon in Baddeck. We need you here. And if you ever run across another bright young student who might be able to help Dr. Bell in his flying experiments, bring him along with you."

The request was something which McCurdy never really thought about seriously until June of 1906 when he was packing up to make the summer trip from Toronto to Baddeck. When he was in the middle of the task, Casey Baldwin idled into the room. McCurdy was never sure afterward whether he had come in out of sheer curiosity, or to say good-bye to him. He merely had the impression that the visit was a very casual one.

"And where are you going to spend the summer?" McCurdy asked. Planning of any kind was seldom a major worry with Casey Baldwin. He was open for suggestions.

And at this moment, McCurdy remembered the invitation which Mrs. Bell had given in a letter quite some time before. "Why not come down to Baddeck with me?" he asked his pal.

"Baddeck? Where is Baddeck?" It is possible that Casey was joking. It is equally possible that he wasn't. Geography may have been one of the several subjects which Casey never allowed to trouble him much.

In any case, Baldwin decided he would go down to Baddeck for a couple of weeks and look the place over. He stayed for forty years.

This is perhaps as good a place as any to take time out to give Mrs. Bell a long-denied credit for her part in the flight story. Genius though he was, Dr. Bell was inclined to be too much of a dreamer to be practical. For all his attention to meticulous detail, he was not even a good mechanic. His wife once accused him of being a man "who could never really finish anything." Once she said: "He would still be tinkering with that telephone if they hadn't taken it away from him."

Nor was he a good business man, and it was his wife who compensated for that. It was she who really managed the immense estate at Beinn Bhreagh and she did the job thoroughly. Dr. Gilbert Grosvenor, Chairman of the National Geographic Society, and son-in-law of the Bells, put it this way: "Mrs. Bell was as great a woman as Dr. Bell was a man. In spite of her handicap of deafness, she was exceedingly well informed on most scientific matters of the day. And though history does not seem to have paid much attention to the fact, her creative talent seemed at times to be scarcely less brilliant than that of Bell himself."

It was her ability as a practical organizer plus her creative talent which gave birth to that unique idea known as the "Aerial Experiment Association."

McCurdy remembered the incident well. "It was on a windy miserable night in September, 1907, and we were all in the big living-room of Beinn Bhreagh warming our backs against the fireplace and talking over our kite flights of that day. Mrs. Bell came in with some hot coffee, watched the conversation for a few moments and said: 'Now Alex, you have some pretty smart young engineers here. And they're just as interested in flight as you are. Why don't we form an organization?'"

Forceful man though he could be at times, Bell had the habit of listening well to his wife, particularly when they were within the confines of the big house on the hill. He turned the matter over in his mind briefly, and within an hour or so an enthusiastic plan began to take definite shape. Not only was the original suggestion that of Mrs. Bell, but it was she who insisted upon financing it. Bell was certainly in no financial troubles at the time, but his wife had come into some money recently from the sale of a parcel of land that she had inherited, and from this she took $20,000 and put it at the disposal of the infant corporation. Later she was to add to that amount.

Members of the group called themselves "associates," and were ultimately five in number. But on its birth night, that gusty September night in 1907, only three of these were present. They were Bell himself, Casey Baldwin, and McCurdy. But Bell already had his eye on two other brilliant young minds of the day. While the Canadian government was at that time utterly sceptical of the practicability of flying and refused to be interested in the experiments, the United States War Department was keenly aware of their possibilities. The United States Army had in fact already been making a few studies of its own. Lieutenant Thomas Selfridge of the First U.S. Field Artillery was their leading aeronautical expert at the time, and Bell's keen eyes had spotted him some months ago. Next to Bell himself, Selfridge probably knew more about the flying of kites than any man in the world, and Bell went directly to President Theodore Roosevelt, asking that the young Artillery lieutenant be granted indefinite leave of absence to aid them in their experiments at Baddeck. Roosevelt released Selfridge at once and without reservation—a fine testimony indeed to the esteem in which the great inventor was then held.

The fifth member of the group was Glenn H. Curtiss, whom Bell had met at a New York motor show. Curtiss, though a very young man then, had already achieved much fame as a manufacturer of motorcycle engines, and he was employing eighty-five men at his factory in Hammondsport, New York. Bell knew that such a man would be invaluable to the newborn association. Motor experts of any kind were few and far between in those days, but Curtiss' particular kind of engine, one which was light, compact and simple, was precisely the kind of power plant which heavier-than-air flight demanded.

October of that year brought Selfridge and Curtiss both to Canada, and the Aerial Experiment Association was formally organized. Bell worked out the details as carefully as if he were concluding a laboratory experiment, and incorporated the Association in both the United States and Canada. Dr. Bell was to be Chairman, Glenn H. Curtiss, director of experiments, Casey Baldwin, chief engineer, J.A.D. McCurdy, treasurer, and Lieutenant Selfridge was assigned the duties of secretary. The Association had as its aim simply: "To get a man into the air."

Bell insisted that no detail of the Association work would be too small to record, and required that once a week the group should publish a bulletin reviewing their activities and accomplishments for the week just past. The Association was to continue for a period of one year. Glenn Curtiss, who came to the group already established as a motor manufacturer, was to receive the sum of $5,000 for taking temporary leave of his factory to come to Baddeck. Baldwin and McCurdy received $1,000 a year each. Selfridge declined any salary since he was already on the full pay of the United States Army.

It was quite natural that the first experiments carried out by the Associates should be with kites. Bell had been flying kites on the hills of Baddeck when McCurdy was a barefooted youngster here, and now nearly twenty years later he was still flying them. Only they were much larger now, and there was much more science hidden in their peculiar skeletons.

Bell placed tremendous faith in the tetrahedral cell as a building unit. He arrived at this design by trial and error after experimenting with every conceivable shape in the geometry book. And now that he had discovered it to be a means of achieving the maximum structural strength with the minimum of weight and materials, he immediately planned to use it to manufacture a gigantic man-carrying kite. But it wasn't really until he had the help of the associates that he was finally able to manufacture a kite which could be depended upon to carry a man aloft.

This is not to say that none of Bell's kites had succeeded in lifting a man heavenward before this time. Some Baddeckers still chuckle about the adventure of Neil MacDermid—the same one mentioned in Douglas' diary—who endured a totally unscheduled flight nearly ten years prior to the organization of the A.E.A. Neil was holding a great kite for Dr. Bell one day when a sudden gust of wind lifted both him and the kite to a height of some 40 feet before he had time to realize what was happening to him. Bell, meanwhile, was running around below in the utmost excitement and shouting into the gale: "Hang on, Neil! Hang on!"

The newly formed Association now took its cue from Dr. Bell and decided that its first project would be to attempt a man-carrying flight with the latest and greatest of all the Bell tetrahedral kites, an enormous and cumbersome affair graced by the name of "The Cygnet." Description of this strange craft almost defies the use of words. But to some extent at least, "The Cygnet" resembled a giant slice of honeycomb, with each of its red silk cells being triangular rather than hexagonal. It was hoped that ultimately she would carry both a man and a motor. And on December 6th, Lieutenant Selfridge was given the chance of taking her up on her first flight.

A lot of genius had undoubtedly gone into her cells, but the fact that she was supposed to carry a man seems to have been all but overlooked. There was no cockpit—merely a longitudinal, culvert-like hole where a man could crawl and lie face down. When a gust of wind struck the craft, the pilot was supposed to shift his weight so that the whole thing would come back to an even keel. If the nose pointed down, her pilot was supposed to shift his weight backward to point her nose up again. Not only was the method of operation rather weak in theory, but when the pilot had crawled into his cramped quarters, his vision was shut off in practically every direction except straight ahead. He was literally buried in hundreds of red silk cells, and had to plot his course by guess as much as by sight.

The plan devised by the Association to launch their kite into the air was relatively simple. First they secured a flat-bottomed, barge-like boat known as a catamaran, and hoisted "The Cygnet" onto it. A local steamer named the *Blue Hill* had been hired to do the towing, and a rope ran from a block on the steamer into the middle of "The Cygnet." The steamer headed into the wind. Selfridge crawled into his uncomfortable manhole, and Bell gave the signal to move off. At a speed of about twelve miles an hour, the great kite took to the air while all hands on board roared their excited approval. In less than a minute she had soared to a carefully measured height of 168 feet. And there she floated with the poise and confidence of a lazy gull.

"The Cygnet" was so utterly placid during the early part of her flight that Selfridge, flat on his stomach and smothered with flapping silk, did not notice when the kite began to gradually lose altitude. One of the sailors on board the *Blue Hill* had been posted beside the rope and was ready to sever it with a stroke of his axe should an emergency develop. But one could hardly expect an ordinary seaman of that time to know much about the emergencies of flying. Furthermore, when "The Cygnet" began to make a deliberate dive for the water, a cloud of black smoke blew from the steamer's funnel and blotted her out of sight. Even after "The Cygnet" had struck the water, the *Blue Hill* raced merrily onward, splintering the fragile craft in the choppy water. Selfridge had barely time to dive clear, but when rescuers rushed to pull him out of the frigid water, he seemed far more concerned over the plight of "The Cygnet" than his own.

"The Cygnet" was a total wreck and beyond all repair. The only fortunate incident of the whole adventure was that Selfridge had taken off his boots before getting into the kite, and had nothing on his feet but long woollen socks. Swimming in oilskins was obstacle enough. The added impediment of boots might have made swimming impossible.

"The Cygnet" was the most famous of all the Bell kites. But she was not quite the last.

7

Aerial Experiment Association Hammondsport, New York, Inc.

Shortly after the disastrous plunge of "The Cygnet" into the icy waters of Baddeck Bay, the associates closed up shop for the winter so far as Baddeck was concerned, and shifted the scene of their operations to Glenn Curtiss' home town of Hammondsport, N.Y. And here the group brought its first joint project into shape. This was the so-called "Hammondsport Glider." The New York hillsides were not only of the right size and shape for gliding, but they were well padded with snow. And the first gliders were so crude that a soft landing place was decidedly important. A light biplane structure was designed which fitted over the shoulders but allowed the experimenter's feet to remain free for running. The would-be birdman would struggle to the top of a high, steep hill and would come racing down the slope in the mad hope that he would get up sufficient speed so that his crude wings would lift him above the brush and stones at his feet. Or if the hill dropped sheer enough, the operator might leap into the air, hugging his knees to his belly and hoping fervently that his invention was at least good enough to keep his hind end from bouncing on the rocks of the hillside.

This crude glider was sadly lacking in stability. It did get a man into the air on more than one occasion, however, but the flight which it offered was far from a gentle one. The first wings had a tendency to undulate and shimmy very much like a wash on a windy day. The associates were not long in adding a tail to the structure, however, and now for the first time the glider had a considerable degree of stability. Flights of thirty to forty yards were not uncommon now, and McCurdy recorded at least one flight of over one hundred yards.

The associates studied the action of their flimsy glider carefully, and set down their observations in detail in the once-a-week bulletins which Bell insisted that they publish. And the experience which they received from these early gliders gave them an insight into the basic problems of flight which was soon to prove its worth.

It was inevitable now that the inventors should think of putting one of Curtiss' efficient little motorcycle engines into a glider. The result was their first motor-driven machine, a gaudy and cumbersome fledgling which they called the *Red Wing*. She was a biplane, for the young engineers had decided that the two-winged type of craft gave the best combination of lift and stability. The *Red Wing* was equipped with a fixed stabilizer, she had a rudder at the rear, and an elevator mounted in her nose. The skeletons of her wings were of bamboo and these were covered with red silk which had been brought down from Baddeck—the same gay fabric which Bell had bought in enormous quantities to make into his famous tetrahedral cells. The craft had an undercarriage of two runners, and a V-eight air-cooled engine which Curtiss had designed especially for the job and which was supposed to have a horsepower rating of forty. Engine rating was apt to be more optimistic than exact in those days, however, and the probability is that the *Red Wing's* engine had an actual horsepower strength of not more than about eighteen or twenty.

It was a little embarrassing to these young adventurers in later years, that the world should want to be so exactly informed as to who actually was the associate who flew first. From the day of its inception, the group had been a perfect team, each admittedly quite dependent upon all of the rest, and there had certainly been no attempt for any one of them to get ahead of his mates. So on the memorable event of March 12th, 1908, when the *Red Wing* was trundled out on the ice of Lake Keuka, near Hammondsport, the fact that it was Casey Baldwin who sat in the cockpit had nothing at all to do with his rank, or some secret ambition of his to be first. The reason was that the associates present, except Casey, had got themselves on skates for the occasion. But Casey, with characteristic good-natured indifference, had simply not got around to finding his, and when the little party finally got out on the lake with the *Red Wing*, Casey was slipping all over the place. So he was no good anywhere else but in the plane itself. The firm-footed members of the party were needed to steady the wing tips and to wind up the motor.

It is interesting to note that while the Wright brothers had been flying for some years at Kitty Hawk in North Carolina, all of their activities were carefully hidden from the public. Bell, on the other hand, believed that he had nothing to hide from the world and spectators were always welcome at any of his outdoor experiments. The projected flight of the *Red Wing* was unique because, if successful, it would be the first public flight of a heavier-than-air machine in North America. So there were several hundred excited spectators on hand that morning to see the *Red Wing* make her attempt to climb into the air.

The lake was frozen solid, and without the slightest trace of snow on its surface. Only three of the associates were present, along with several of the factory hands from Glenn Curtiss' motorcycle works. Bell and Selfridge were both in Washington at the time. All the men there were on skates. All but Casey.

Disaster almost overtook the *Red Wing* the moment the motor started, because through some momentary confusion, Casey was not in the cockpit at the time the propeller was spun. So the plane started off across the ice, the master of its own fate, and actually rose a foot or two into the air before settling back again. No damage was done and the skate-shod pursuers soon caught up to her, hung on to her tail, and finally nosed her back to the original starting point.

This time, Baldwin was ready in his seat, Curtiss gave a few last minute love touches to his motor, and then coaxed her into her top speed while the helpers placed themselves at the wing-tips to hold the craft back.

Finally Baldwin signalled that he was ready, tested the controls, and hollered: "Let's go!" The men who had anchored the wings sprang back, the little *Red Wing* jumped forward, picked up speed rapidly, and in a few seconds she left the ice. The actual distance which she taxied before taking off was measured later and found to be 150 feet. As soon as he felt his craft free of the earth, Baldwin pulled her nose up rather sharply and she climbed into the air like a scared bird to a height which varied from four to five feet. And she held this height for a distance of 319 feet. Then the motor, as though she had accomplished the most tremendous job the world had ever seen, gave up her gallant struggle and stopped dead. Baldwin made a perfect landing nevertheless.

In this regard he was very fortunate, though the associates did not realize it at the time. There was not a trace of wind that morning and for that reason, there was nothing to disturb the *Red Wing's* stability. Had the *Red Wing* been struck by a gust of wind, particularly after her engine had gone dead, she would doubtless have fallen into a splintered heap on the ice. In fact, that is exactly what happened just five days later, for on March 17th, St. Patrick's Day, Casey was again attempting to fly the *Red Wing*, but this time he had to deal with a slight wind. The wind came under one of her wings, lifted it sharply, and since there was as yet no device to compensate for such an emergency, the *Red Wing* upset on the ice, completely destroying its wings and wrecking its motor as well.

The flight of the *Red Wing* was given tremendous publicity in the newspapers of the day, but many of them made a rather unforgivable mistake in reporting the incident. The name of Casey Baldwin was practically unknown to them, and the press gave the credit to a Captain Baldwin who was well noted in those days for being an outstanding balloonist. He was also a friend of Dr. Bell and a frequent visitor at both Baddeck and Hammondsport.

To the associates, the death of the *Red Wing* was of greater importance than her short but glorious life. For one of the party had taken a picture of her a split second before the disaster occurred and by studying this photograph, the associates knew very well what problem must be solved next before any flying craft could achieve stability. Once the urge overtook the plane to point one of her wings at the sky and the other at the earth it was impossible for her pilot to level her out again. And as long as there was the slightest trace of wind, emergencies of this kind had to be anticipated. The Wrights too were having the same difficulty, although the associates could not have known it at that particular time. But it was later discovered that the Wrights tried to stabilize their flights by a device which twisted the whole wing structure so that the wings would lift on one side of the craft and dip on the other at the command of the pilot. This "warping" device had a very bad fault, however, for it relied upon a single wire to preserve the strength of the structure. It was later abandoned as structurally unsound.

The development of the aileron was entirely that of the associates, and was one of the really great aeronautical advances of the age. In its earliest form, the aileron was simply a flat flap mounted at each wing tip, and these were attached to the body of the pilot by a harness. The theory was ridiculously uncomplicated. When the flyer felt a wing beginning to lower on his right, for instance, it would be his instinctive tendency to lean in the opposite direction. Now, with his lean harnessed to the aileron, the wavering craft got the corrective action it so desperately needed.

The naming of the new device was spontaneous and accidental. McCurdy was visiting the famous French experimenter Farman many months later and he, on learning of this strange new apparatus, exclaimed excitedly in French: "Why, it's a little wing!"

Hence the French origin of the name "aileron," which is now an important word in our language.

One other device went into the making of the next craft which came out of the associates' Hammondsport shop that winter. This was the tricycle landing gear. The wheels were light motorcycle wheels, borrowed from the Curtiss factory. Previous to this, all experimental flying craft had used either runners or a launching device such as the Wrights had at Kitty Hawk.

That next brain child of the group was called the *White Wing*. The supply of the red kite silk from Baddeck was exhausted now, and the laminated ribs of this next craft were covered with a white soft cotton called nainsook.

The motor which had lifted the *Red Wing* had passed into oblivion with the plane itself, and Curtiss refused to mourn its loss. "It was no good anyway," he said. And in truth he already had a better engine in the process of construction.

Curtiss rushed the engine to completion, while the other members of the group were building the *White Wing*. It took nearly two months to complete the new craft. The designs were drawn mainly by Baldwin though the actual manufacturing was in the charge of William F. Bedwin, who had been superintendent of Dr. Bell's kite plant back in Baddeck.

On May 13th, 1908, the *White Wing* was ready to commence her memorable part in the history of aviation. There was no smooth sheet of ice on the lake for her launching this time, and the Hammondsport half-mile race track had been selected as the best flying field available.

Now a rather serious problem developed, for it was discovered that when the *White Wing* ran along the smooth surface of the race track, her outer wing was running afoul of the fence which circled the track. There was a rather sharp diversity of opinion as to what should be done now, with the Canadians suggesting recklessly that the fence should be torn down. Selfridge and Curtiss were more discreet and thought that the plane might very well be flown from the centre of the oval rather than from the track. Bell, always a great internationalist, gave the deciding vote in favour of the Americans. But when the *White Wing* attempted to make a run on the interior of the oval, it was found that the ground was so rough the wheels began to buckle. A solution was finally found by going to Curtiss' again for heavier wheels, and with her undercarriage thus reinforced, the *White Wing* was given a chance to prove her worth.

Selfridge tried first. He did not try to get off the ground, however, but contented himself steering the new craft around the field and getting the feel of her controls. Once again Baldwin was the man who got off the ground first and this he did on May 18th, just five days after the day when the *White Wing's* first wheels came to grief. He flew for a distance of 93 yards, again at the height of ten feet, but a slight accident prevented him from extending the *White Wing* to her utmost that day.

When he came down, Baldwin said, "I'm sure she would have flown for several minutes, but the elastic rear edge of the lower wing was fouling my propeller. So I had to come down."

Apparently the pressure of the wind on the wing was more than the designers had anticipated, but little damage had been done and it was easily repaired.

It is interesting to note that comparatively little thought was given in those days to balancing a propeller with any great degree of care. And fouling a propeller seems to have been as common to the early flyers as flat tires were to the early motorists. Says Selfridge in his part of the "Bulletin" which describes the flight which he made in the *White Wing* immediately after Baldwin's attempt: "Loose wires fouled the propeller."

It was Glenn Curtiss who first flew the *White Wing* with marked success, and on his first flight he established a new world's record with a flight of 339 yards. It took the *White Wing* just nineteen seconds to cover that distance, so he was flying at about thiry-seven miles per hour.

In his section of the "Bulletin" report, Curtiss describes his flight as follows:

"As usual, half a dozen workmen held on to the wings until the motor was developing her full power. Then they let go, and the plane shot down the field for a distance of about 150 feet. I pulled back on the elevator control and she sailed into the air. She seemed to me to be climbing too steeply, however, so I pushed her nose down sharply. As is usually the case with an inexperienced operator, I overdid it, and had to again pull back sharply to keep from smashing into the ground. I found out afterward that the wheels touched the ground in the middle of the flight. However, she climbed to a height of about ten feet and ran beautifully for another 134 yards. In the first stage of the flight, the plane travelled 205 yards.

"At all times, the controls worked perfectly and the plane answered without a hitch, and in my opinion all that holds us back from much longer flights now is the inexperience of the operator. That will, of course, come in time, and when it does we will be making flights of an hour or more."

Dr. Bell never flew. And of the four remaining associates, McCurdy was the only one who had still not got into the air. This was a rather natural happening because McCurdy was the youngest of the four, but it was his turn now.

However, his first flight was very nearly his last, and in it he not only came near to killing himself but Selfridge as well. When McCurdy climbed into the cockpit of the *White Wing*, Selfridge walked some distance down the track in front of the plane and took up a position there with his dog, so that he could note the precise time the machine left the ground. It was his intention to mark this spot for future reference. But once McCurdy set off in the *White Wing*, he urged so much speed out of his craft that Selfridge suddenly came to the frightful realization that he could not get out of the way in time. He immediately flattened himself on the ground and barely in time to have the *White Wing* pass over him harmlessly. McCurdy, meanwhile, was so intent upon the perilous thrill of his first flight that he didn't even notice Selfridge. He recalled later, rather vaguely, that he had seen a dog scurry off into the grass, but he was quite surprised to know that he had barely missed his partner.

The flight itself was successful enough, for McCurdy lifted the plane to a height of twenty feet and kept it there for 183 yards. But his landing was a complete failure. In the words of one of the frightened observers present, "He failed to correct with his lateral controls the effect of the quartering wind which was blowing." The result was that he tilted over on his right wing and struck the earth in a tremendous cloud of dust and splintered framework. The crowd rushed on the field fully expecting to find McCurdy a mangled corpse. But he was quite alive and when they reached him, he was sitting on the ground a few yards away from the plane calmly examining himself for broken bones.

McCurdy himself never agreed that the crash was really a dangerous one. In an article in the "Bulletin" he described it thus:

"It was a comparatively calm day, the wind only coming in puffs, but it was through one of these puffs that the machine met its Waterloo. Curtiss started the engine and, as in previous trials, the plane was held by a half dozen men till the engine was developing its full power. She then ran along the ground and left it so smoothly that I didn't realize we were in the air.

"The machine took a slight turn to the left and then curved round to the right. The wind blew about on the left quarter, and as she turned to the right, another puff elevated the left wing, depressing the right so much that it caught in the grass. I leaned to the high side with the idea of adjusting the tips so that a righting could be produced, but as I was sitting too far forward, my back failed to engage the lever which operates the tips and so no righting result was produced.

"As the right wing struck the ground, the machine pivoted and the nose swung around and dug into the ground. I was deposited gently and without much of a jar on the ground. The machine turned a complete somersault almost over me, but left me free of the debris."

Whether McCurdy agreed or not about the seriousness of the crash, it meant the end of the *White Wing*. She was so thoroughly broken that nothing could be salvaged from her but the motor.

Proudly convinced that the young birdmen whom he had fathered would go on strengthening their wings without his help, Bell spent less and less time at Hammondsport. He had other duties in Washington, and Toronto. Already the plans of the *June Bug* were nearing completion on the drawing board. In this craft

Curtiss took the lead in designing the plane as well as the motor which powered her. And in her earlier flying days, he was the one who piloted her most often. The creation of the *June Bug* brought still another important contribution from the Aerial Experiment Association. One morning after she had proven herself several times, the *June Bug* seemed to be very reluctant to take the air though her motor was functioning perfectly. McCurdy noticed that the fabric of the craft was very heavy with dew and rightly concluded that the added weight was the reason for her laziness. In a remarkably short time, the first aeroplane "dope" was applied to her wings. This was a concoction which they made up of paraffin, ochre and gasoline.

The *June Bug* never crashed and her performance was so universally satisfactory and Curtiss so adept at navigating her that the associates decided to make a bid for the so-called "*Scientific American* cup" which was the first official trophy to be offered for heavier-than-air flight. The rules of the test were that in order to qualify as a challenger for the cup, the machine must fly a distance of one kilometre in a straight line. Such an accomplishment now seemed remarkably easy to the associates, and one day with their spirits soaring as high as the new plane, the following wire was sent to Dr. Bell, who was thought to be somewhere between Toronto and Baddeck:

"TO ALEXANDER GRAHAM BELL, PRESCOTT, ONTARIO, OR KINGSTON, B/O STEAMBOAT TORONTO OF B.-O. STEAMBOAT LINE, WHICH LEFT TORONTO TODAY AT 2 P.M. FOR MONTREAL, AND IF TOO LATE TO CATCH BOAT REPEAT TO WINDSOR HOTEL, MONTREAL, QUE.

"HAMMONDSPORT, NEW YORK, JUNE 25, 1908. CURTISS FLEW 1140 YARDS IN SIXTY SECONDS THIS EVENING ABOUT 7.30. WE HAVE TELEGRAPHED AND TELEPHONED SECRETARY AERO CLUB OF AMERICA THAT WE ARE NOW READY TO TRY FOR THE SCIENTIFIC AMERICAN CUP. HURRAH!

"— SELFRIDGE."

The date for the challenge was set for Independence Day, July 4, 1908, and more than a thousand spectators flocked to Hammondsport. Other than the fact that the *June Bug* was the first heavier-than-air craft to have its name inscribed on the *Scientific American* trophy, there was nothing unusual about the flight itself. Curtiss, in a rather laconic entry in the "Bulletin," stated: "I could have continued the flight after the measured mile had I wanted to, but to do this would have meant flying over the trees which surrounded the field, and I did not deem it wise to do this at the present stage of my flying development."

Not all of the spectators at Hammondsport that day were holiday travellers and thrill seekers. Evidently there were many other experimenters mingled in the crowd and some of these did their best to decry the importance of the flight. An interesting revelation of this fact is found in the letter from Mrs. David Fairchild, the younger of Dr. Bell's two daughters, to her father:

"In spite of all that I had read and heard, and all the photographs I had seen, the actual sight of a man flying through the air was thrilling to a degree that I can't express.

"We all lost our heads and shouted and I cried and everybody cheered and clapped, and engines tooted. Mr. Post was there, and Mr. Hawley, Vice-President of the Aero Club, Mr. Herring, Captain Baldwin the balloonist, Mr. Lake of Submarine interests, and Mr.—— a very nasty, grumpy individual who, however, was not able to interfere with anyone's enthusiasm. All sorts of pictures were taken, and the air was full of the click, click of shutters. There were moving picture cameras and kodaks of all sizes.

"At the first flight I was at the corner of the road nearest the barnyard with Douglas [McCurdy] and David [Mrs. Fairchild's husband] was at the starting line. The machine rose beautifully and flew by us but didn't quite make the mile. It was flying pretty high and Mr. Curtiss wanted to bring her down a little, but she answered her controls quickly, and when he got her down he could not get her up again. Nothing was hurt, and all hands towed her back to the starting point for the second attempt. Mrs. Curtiss and I chose our stand on an old log on the far side of the potato patch. The first flight had raised excitement to the boiling point, and as Mr. Curtiss flew over the red flag which marked the finish and went on his way toward the trees, I don't think any of us knew quite what we were doing. One lady was so absorbed as not to hear a train coming and was struck by the engine and had two ribs broken."

It was with the *June Bug* that McCurdy first achieved the status of a great flyer. He flew the craft often and made several notable records with it. Perhaps the most notable of his accomplishments was the flying of the first figure eight. Such a feat had an importance far beyond its novelty, for it demonstrated in a striking fashion the manoeuverability which was now possible through the use of the ailerons devised by the associates.

McCurdy was so thrilled with what he had done that he sat down that evening to write Mrs. Bell about it:

"Last evening, about five o'clock, I attempted a flight and much to the pleasure of everybody, I was able to describe the figure eight, after covering about a distance of two miles. I took off and made a turn successfully, fully intending to land when I got back to my starting place. The engine sounded so well, however, and the people were excited and yelling, so I decided I might as well try another turn. I could see the people with their mouths going, although I couldn't hear anything on account of the noise of the engine. I am becoming more competent with each flight, and, although I am afraid you will think I am boasting, feel sure I will soon be able to fly anywhere."

The patents for the aileron were later given to the United States government by Bell during the First World War. It was a noble gesture indeed and one which might be expected from so public-spirited a citizen. But had the associates retained the patents for themselves, they undoubtedly would have earned millions.

The *June Bug* had a strange change of name during that summer, for McCurdy decided that he would like to transform her into a seaplane and fitted two light pontoons to her belly. He now called her the *Loon*, a name which proved most fitting because the loon is a water fowl which needs a very broad sweep of open water before its take-off will lift it into the air. With the Curtiss *Loon*, however, no amount of open water and no amount of motor power would lift her into the air. The associates discovered later that the motion of the pontoons over the water created a suction which held the craft fast. She could go forward, but not up. Their invention of a "step" in a pontoon in 1911 was the solution for this difficulty. But that invention did not come in time to make the *Loon* airborne. She never got into the air until she traded her pontoons for the original motorcycle wheels.

Just as the associates were rapidly becoming known as the world's most successful flyers, stark tragedy struck. In the early fall of 1908, the U.S. War Department contacted Lieutenant Selfridge, asking him if he would act as a government observer in a public flight which was to be made by the Wrights near Washington, D.C. There was no thought of sadness when Selfridge took his leave of Hammondsport and his faithful partners. For all of them had wondered for a long time what the Wright machine must be like and now they would know for certain.

But Selfridge never came back. He was instantly killed in the crash of the Wright machine, the first man in the world to be sacrificed in the battle for mastery of the air. He was given a military funeral, and Selfridge Field, near Detroit, Michigan, still honours his name. But when the news of his passing got back to Hammondsport, a great gloom settled over the little shop and for a time the associates did not know whether they should go on with their wild idea or not.

In one of the "Bulletins," the death of Selfridge is set down, not by an associate, but by a man named Clime, who happened to be an eyewitness to the tragic event:

"For several complete circuits of the field the flight was uneventful. The novelty had worn off to such an extent that one no longer kept his eyes glued to the machine but only gave an upward glance when it went directly overhead. It was at such a time that Wright (Orville) could be seen, hands on levers, looking straight ahead, and Lieutenant Selfridge to his right, arms folded and as cool as the daring aviator beside him.

"While walking over to the aerodrome shed, there was a crack like a pistol shot coming from the air. Looking quickly up, I saw a piece of propeller blade whirring off to the southward. Realizing instinctively that something terrible was about to happen, I stood riveted to the spot with my eyes on the machine. For a brief period it kept on its course, then swerved to the left and took a swoop backward, then in an almost perpendicular manner it fell for half the distance to the ground. Suddenly righting itself, it regained for an instant its normal position, only to pitch forward and strike the ground, raising an immense cloud of dust which momentarily hid it from view.

"The terrific impact reduced the structure into an inconceivable mass of wreckage. At topmost speed I ran over to where the machine lay, and found that two mounted soldiers had preceded me by a few seconds. I threw my camera down, caught hold of one of the curved surfaces, and with all my strength pushed it up and broke it. Mr. Wright was under it and only a few feet from me, apparently in great pain, and moaning. He had fallen across a wire stay and one of the struts of the engine, and was suspended by his chest and stomach. His feet were barely touching the ground and his hands were hanging limp. Blood was streaming down his face and trickling in a stream from his chin, but he was conscious and feebly said, 'Help me.'

"Lieutenant Selfridge was lying on his back almost directly beneath Wright but a little to the right. One of the enlisted men reached over, caught Mr. Wright around the body and lifted him clear of the wreckage. It was then that he exclaimed, 'Be careful of my leg!' The two enlisted men and myself lifted and carried Mr. Wright away from the debris, and then turned back to the wrecked plane. We tried our utmost to reach Lieutenant Selfridge, who was lying on his back and had apparently struck the ground with the back of his head and the base of his spine. Wire stays, pieces of canvas and broken struts were piled in confusion around him, and it was impossible for the few at hand to do more than feebly attempt to reach him. He was unconscious, and if he spoke at all, I did not hear him."

Back in Baddeck, the sad news reached the Bells. And knowing how profoundly Selfridge's death would affect the rest of her boys, Mrs. Bell sat down to write them a few words of comfort. The letter is dated September 19th, 1908, just two days after the crash:

"I can't get over Tom's being taken. I can't realize it; it doesn't seem possible. Isn't it heartbreaking? And yet it is better for him than to die as poor Langley did. [Langley died broken-hearted over lack of success and ridicule.] He was so happy to the very end. I know he would have said he was having the time of his life, and, though he must have really realized his danger in those last few seconds, he would still hope to escape, and he had no time for unavailing regrets. It was the happiest way death could have come to him now, but why need it have come now when he was ready to put to his country's use all the results of his long, patient preparations? . . .

"I am so sorry for you in this breaking of your beautiful association. But it was beautiful and the memory of it will endure—'Bell, Curtiss, Baldwin, Selfridge and McCurdy.' It was indeed a 'brilliant coterie,' as one newspaper said. Do anything you think best, but let the A.E.A. be only these to the end, and then take some other name."

There wasn't much more flying at Hammondsport that summer, but for McCurdy at least, the siren call of the heavens was still shrill and insistent. Already he had brought the A.E.A.'s fourth plane out of the shops, and had coaxed her to prove her wings over Hammondsport. But he had been away in the States for a long time now, and the hills of home were calling. In October of that year, he and Curtiss dismantled the *Loon* and put her parts into crates. The completed parts of the latest plane were also crated for shipment. And just to prove that the folks back home had been following the fame of their favourite son, a petition had gone out from Baddeck to the Government of Canada, asking that the aeroplanes be allowed into the country without paying duty. Duty on luxury items coming into Canada has always been high, and the newfangled aeroplane would most certainly be considered a luxury.

The government gave the matter due consideration, and finally decided to allow the two planes to enter the country duty free with one provision. They had to be taken out again within a year's time.

The customs men smiled wisely about the special condition. "If them there planes ever leave the country again, it will be as match wood," one of them said.

8

The Great Day

It had been Bell's intention to have each of the four young engineers have a turn at designing a plane—or as Bell insisted upon calling flying craft at that time—an aerodrome. The *Red Wing* had been designed mainly by Selfridge, though it was Baldwin who flew her and who had sent her to an early grave. The *White Wing* was of Baldwin's design; the *June Bug* was the proud brainchild of Glenn Curtiss. Now it was time for McCurdy's plane to take shape and, if good enough, to take the air.

There was no doubt now that she was good enough. McCurdy had studied the faults of her predecessors carefully and set about to avoid these wherever possible. Perhaps the greatest improvement of all was in the motor.

While it was true that the *June Bug* flew at least 150 successful flights in Hammondsport, there was at least once when she gave the associates an anxious moment. One day after a good flight of two minutes and twenty-three seconds, her motor stopped. When she came down, and the associates had time to look her over, the reason was plain enough.

"She's frozen," McCurdy said. "She's frozen with heat."

So when he began drafting the plans of the plane which later became famous as the *Silver Dart*, he got Curtiss' help in preventing such a thing from happening again. The result was the world's first water-cooled aircraft engine, with copper-jacketed cylinders.

The only thing which McCurdy gave no great thought to about his machine was her name. He seems to have named her quickly and rather casually. As he told Mrs. Bell later, "The silver part was easy enough. That's the colour of her from the new dope we put on her wings. And the dart part—well, it just seemed natural."

An interesting fact about "the silver part" was that the man who was responsible for it was the Captain Baldwin, the balloonist, who had been wrongly reported as the man who had flown the *Red Wing* in March of that same year. Baldwin was with the Goodyear people and his preparation for plane wings was a silver-coloured rubber compound. It worked admirably.

The *Silver Dart* flew several times in Hammondsport and her longest flight seems to have been for a distance of about one-and-a-half miles. So McCurdy was already well acquainted with her moods and possibilities when he reassembled her in the Bell shops at Baddeck that next winter.

In the same shops, Dr. Bell was making one final attempt to put a motorized kite into the air. Officially, Bell recognized this latest kite as "Drome V." She was still a tetrahedral affair and it was agreed that the new water-cooled engine, which Curtiss had designed for the *Silver Dart*, might first be tested in the kite. The engine arrived on February 11th at Beinn Bhreagh and was installed February 22nd. Bell lost no time in fitting her to his kite, but his attempts at flight were unsuccessful and he broke a valuable propeller to boot.

He was still not quite ready to give up the dream of a great motorized kite, however, and ordered another propeller. While waiting for this one to be made, the new Curtiss engine was transferred to the *Silver Dart* on the morning of February 23rd, 1909. And before noon, the news ran like a little trail of fire about the Baddeck community that McCurdy was going to fly his *Silver Dart* over the ice of Baddeck Bay that afternoon.

The whole countryside turned out to see the attempt. Many of the old timers came fully prepared to witness a failure, perhaps a tragic one. They had heard of the flights in Hammondsport, but then the United States was a long way off to them. And even in those times, Maritimers had a quiet distrust of American ballyhoo.

School was let out for the occasion, and most of the pupils had the foresight to come on skates. Among the school children who were given a half-holiday were Clarence and Norman Bethune, younger brothers of the Bethune boys who had witnessed the gunpowder accident fifteen years ago. Clarence later became head of the Victoria General Hospital in Halifax, and his brother Norman ran a garage directly across from the old homestead.

John MacDermid's one-horse sleigh was pressed into service to tow the craft on the ice and McCurdy's cousin Mabel insisted on bundling herself into the sleigh with him. Perhaps she wanted to be handy to render first aid. About eight of Bell's staff pushed the craft from various points at the wings and tail. McCurdy, wearing a stocking cap, walked in front of her a little to the left of the engine. The 42-foot wingspan of the *Silver Dart* was balanced airily upon the usual light motorcycle wheels, though she was later tried with runners as well. McCurdy worked over his machine, testing the engine and the controls for some time and seemed impatient to get the adventure started. Finally, resplendent in a spanking red cutter, and mountainous in his fur coat, Bell himself drove on to the ice. McCurdy told the inventor that he was ready to go now, but Bell held up a restraining hand. "It isn't nice to think of it, but there just might be an accident," he said. "And the doctor isn't here yet. We'll wait for the doctor."

Finally Dr. Dan MacDonald, hurried as always, drove up and apologized for being late. McCurdy headed the *Silver Dart* into the slight breeze, and someone cranked the propeller. He listened to the motor for one last time and it sounded good. He signalled the crowd away and let the new Curtiss motor whip the *Silver Dart* down the ice and finally into the sky. The *Silver Dart* lifted herself to the frightening height of thirty feet and flew for three quarters of a mile at about forty miles per hour. Since McCurdy had intended this to be a test run only, he now turned her around and came back to his original starting place. He made a perfect landing as the crowd cheered and threw hats and mitts into the air.

McCurdy got out of the plane and reported dutifully to Dr.

Bell. "She's all right," he said. "Now I'm going to take her up for a real flight."

But once again, Bell restrained him. He made a rather dignified little speech which was for the crowd as well as for McCurdy. "What we have seen just now may well prove to be one of the really important pages in history," he said. "I wouldn't want it to be spoiled. Douglas, you can fly her again tomorrow if you like, but that's all for today."

It was now time for a celebration, and everyone in the crowd was surprised to learn that Bell was completely prepared for festivities. Bell had always loved theatricals, especially homespun ones, and a few years previously he had rigged up a stage, complete with modern stage curtains, in the loft of his warehouse. He invited the crowd there now, and when the villagers arrived, they found that he had provisioned the place with sandwiches of all descriptions, made in the big house on the hill. There were gallons of steaming coffee and tea as well, and the inevitable raspberry vinegar. Bell had an inordinate love for raspberry vinegar, most of which was made at Beinn Bhreagh.

Now, with the same methodical manner which followed him to his study or laboratories, he set out to record the precise happenings of the day. A book was provided, and everyone who had witnessed the flight was asked to sign his name. There was still another speech and later that night, Bell recalled as much of it as he could and set it down too in the record.

But when it was McCurdy's duty to enter the event in the weekly "Bulletin," there was very little evidence of anything but the technical details. He states simply as follows:

"The morning of today (February 23) was spent in getting the *Silver Dart* ready for a trial flight. The transmission was changed from the four V-belt drive to a single chain drive which, it was anticipated, would not only give greater efficiency, but would be of less weight. The gearing used was 18–24 (or 3–4), the engine turning over 24 revolutions to the 18 revolutions of the propellers.

"We had three propellers. To decide which one was to be used, a series of tests was made on the iceboat, although this was not allowed to advance during any of the tests. The propeller finally decided upon was one having a diameter of seven feet six inches, and a pitch at the tip of 20–22 degrees. This propeller was not one of constant pitch speed.

"The *Silver Dart* was finally taken across the bay on the ice, and a start made at a spot just off Fraser's Pond. In the first trial, a gasoline pipe broke after the machine had travelled about 100 feet.

"Upon fixing this, a second start was made, which was very successful. The machine rose from the ice after travelling about 100 feet, and flew at an elevation of about ten to thirty feet directly east for a distance of about half a mile. Landed without any jar whatsoever. The speed I should judge to be about forty miles an hour."

The newspaper stories, however, were inclined to describe the event with considerably more colour. In *The Washington Star* the next week, Charles R. Cox wrote:

"On Tuesday, February 23, the Aerial Experiment Association continued their experiments. About one o'clock hardly any wind was noticed, and like the previous day the weather was fine and an even greater crowd than the day before had assembled on the ice to witness the experiments. About three o'clock the *Silver Dart* was wheeled from her shed to the outer bay and placed in a position about a mile from the Beinn Bhreagh shore, facing the wind.

"Mr. McCurdy took the aviator's seat. Crowds at this time began to congregate in front of the machine and along the line of advance of the machine until it became obvious that it would be necessary to appoint police in order to keep the ice clear, thereby avoiding accident. Just as the machine was about to be started, the wind shifted from the southeast to the northeast, and it was decided to take the machine farther up the bay and start flying towards the Beinn Bhreagh shore. As most of the laboratory staff were on skates, this was done very quickly, and before some of the people could realize what was taking place, the buzz of the engine could be heard in the distance, and the machine was seen advancing rapidly along the ice. She had gone about 90 feet along the ice when all of a sudden she rose gracefully into the air to an elevation of about 20 to 30 feet, and was travelling at about 40 miles an hour. Every one seemed dumbfounded and before they could realize that they had actually witnessed the first flight of a heavier-than-air machine in Canada, Mr. McCurdy was compelled to shut off his power and glide to the ground owing to the long stretch of land and trees in front of him.

"Just as he was about ten feet from the ground, he noticed two little girls skating in front of the machine, and if it wasn't for his presence of mind, and his complete control of the machine, a serious accident might have occurred. Gracefully did he steer to one side of them, making a beautiful landing on the ice. It was evident to the observers of this experiment that Mr. McCurdy could have flown for an indefinite time as the engine and machine were working beautifully, and it was only a question of how much fuel he had as to how long he would remain in the air.

"Everybody rushed to congratulate this young aviator, and he promised them that he would do better the next day if the weather conditions were good.

"The names of 147 witnesses of McCurdy's first flight in the *Silver Dart* on Baddeck Bay are recorded in the proceedings of the Aerial Experiment Association. No less than 74, or 50 percent of these commenced with the syllable 'Mc' and 16 were 'McDonalds.'

"The following day a larger crowd assembled to witness the flight, and it appeared that the whole country had gone flying machine crazy. McCurdy made a remarkable ascent and flew a distance of four and a half miles, the longest ever made by a member of the association, and described as 'one of the prettiest that has ever been made by any aviator in the world.'"

The distance record at that time was held by Wilbur Wright, who flew seventy-seven-and-one-half miles at Le Mans, France, on December 31st, 1908, in two hours, twenty minutes and twenty-three seconds.

Just to prove that her first performance before the good folk of Baddeck was no mere stroke of luck, McCurdy took her up the next day for another "pretty" flight of four-and-one-half miles. He was also gaining altitude now and began to soar at the dizzy height of sixty feet, a feat which set some of the more conservative heads in the village shaking with grim foreboding. It was no use to carry this taunting of the heavens too far, they thought.

Writing again in *The Washington Star*, Charles Cox described the second flight as follows:

"During his second flight, this pioneer pilot of Baddeck flew a straight course of a mile and then made a wide circle to return along the other shore at an elevation of some 50 feet and at 40 miles an hour. It is recorded that when he was making this turn, the people on the opposite shore with their teams did not know what to do when they saw this 'wizard' coming their way, as some of them exclaimed. One man lost his head completely, and drove directly under the machine, which went over him like a shot. Other people were running for the beach. But away sped the *Silver Dart* over teams, over people's heads, over trees and over a large tongue of land. When making for the lower end of the Bay for his second turn, McCurdy realized that the space was too narrow. He shut off his power and glided to the ice after having flown four-and-one-half miles. In landing, one of the wings brushed the ice and was slightly damaged, while one of the front wheels was bent."

On the third day, with a straightened wheel and the wing tip suitably bandaged, the *Silver Dart* took off for an epoch-making flight of twenty miles over the Baddeck countryside. And on each good day following, she made similar distances without an accident of any kind. On March 8th, 1909, McCurdy covered a distance of twelve miles.

By summer of that year, though the newspapers and the scientific periodicals of the outside world were as loud as ever in their praise of the performance of McCurdy in his *Silver Dart*, Cape Bretoners had outlived the stage when they flocked by the hundreds to Baddeck every time their favourite son decided to try his wings.

In the fall of that year, McCurdy had been relegated to the small print again in the Cape Breton newspapers. The following dispatch, date-lined Sydney, Nova Scotia, June 4th, printed this news in less than an inch of column space:

"Mr. J.A.D. McCurdy made another trial flight near Bentinck's farm, a few miles from Baddeck, making fourteen rounds on a distance of fifteen miles. Both the start and the landing were easily effected. A height of eighty feet was reached and the river crossed four times during the flight."

The Hon. J.A. Douglas McCurdy with his mother, Elsie, and with his father, Arthur, inventor of the Kodak developing tank. Mabel Bell persuaded Arthur to patent that device. Young Douglas became a shareholder by investing the proceeds from the sale of three Shetland ponies. Later, both Mrs. Bell and the McCurdys profited from the sale of that patent. After his mother died, the Bells, who had lost two sons in infancy, had hoped to adopt Douglas. Instead, the three McCurdy brothers—Lucian, Douglas and George—were raised by their aunt, Miss Georgina McCurdy.

In the photograph above, Douglas McCurdy is the small boy holding his hat, standing in front of Mabel, Mrs. Alexander Graham Bell. The photograph was taken by Theodore A. Kiesel at the turning of the sod for the Volta Bureau Building in Washington, D.C., on May 8, 1893. Dr. Bell financed this institution for the study of problems of the deaf, from proceeds of the sale of the wax record patent to Edison. Daughters Elsie May Bell and Marian Hubbard Bell stand to the left of Helen Keller (in dark skirt), who removed the first sod. Dr. Bell's mother, Eliza Grace Symonds Bell, is seated, holding an umbrella. Dr. Bell stands next to Mabel in the centre of the picture, and beside him is Arthur W. McCurdy, Douglas' father. Mrs. David Bell is seated at the extreme left. Behind her is Alexander Melville Bell, the inventor's father, a famous teacher of the deaf. Next to him stands Miss Annie M. Sullivan, Helen Keller's teacher. Right, Douglas, Lucian and George aboard their houseboat in 1898. Douglas McCurdy's diary starts on page 123.

J.A.D. McCurdy with his brothers Lucian and George. Below, the brothers are building the wing of an airplane. The centre photo is from his barnstorming days. And, below right, when he served as Lieutenant-Governor of Nova Scotia, 1947–1952.

Douglas McCurdy grew up in the Baddeck where Alexander Graham Bell, a family friend, was sending an array of odd kites into the sky, experimenting toward controlled, powered, manned flight. Bell had told Watson that after the telephone his next goal was flight, and on his honeymoon he told Mabel that he had dreamed of telephones attached to flying machines. Bell had two firm requirements: stability and safety. He thus invented the tetrahedral cell—see the sketches from his Bathtub Notes—a unit for building a kite that provided strength and relative lightness. Recognizing that his kite experiments might be a dead end, Mabel encouraged and financed the Aerial Experiment Association—four young men who worked with Bell and shared ideas. Together, they produced the "Cygnet," a kite of 3393 cells that carried a man aloft. But when that kite was destroyed in a crash, they decided to experiment along the lines of the Wright brothers' successful biplane. That put them on the road to the *Silver Dart*.

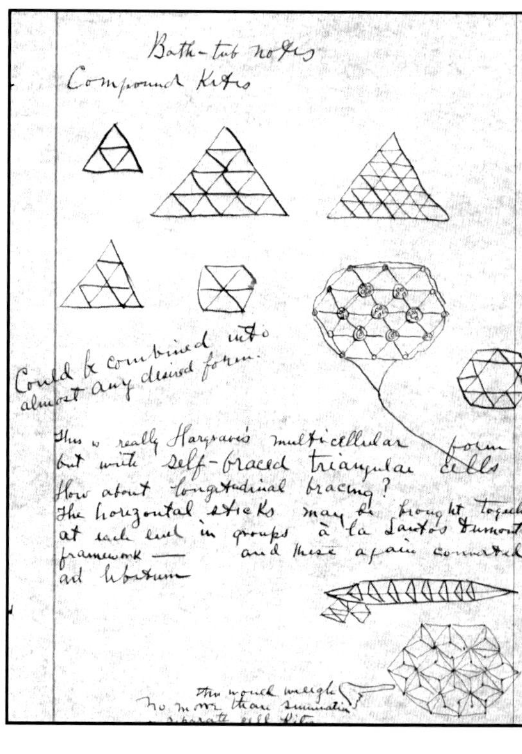
Bath-tub notes
Compound Kites
Could be combined into almost any desired form.
This is really Hargrave's multicellular form but with self-braced triangular cells
How about longitudinal bracing?
The horizontal sticks may be brought together at each end in groups à la Santos Dumont's framework — and these again connected ad libitum
This would weigh no more than summation

Dr. Alexander Graham Bell, assisted by workmen and grandson Melville Bell Grosvenor, hauls down one of the great kites following an experiment. A workman demonstrates the position of the "pilot," who would ride on his stomach in the centre of the Bell kite "Cygnet." Lieut. Selfridge went aloft in that kite at Baddeck and narrowly escaped when the kite, towed by the lake boat *Blue Hill,* plunged into Bras d'Or Lake and was dragged to pieces.

The five members of the Aerial Experiment Association are shown in this photograph taken by Mabel Bell. Left to right are Glenn H. Curtiss, F. W. "Casey" Baldwin, Dr. Alexander Graham Bell, Lieut. Thomas E. Selfridge, and J.A.D. McCurdy.

Early in 1908 members of the Aerial Experiment Association were in Hammondsport, New York, experimenting with a glider. Above, an unidentified member of the group soars from a hillside in February of that year.

On March 17, 1908, the *Red Wing*, piloted by Casey Baldwin, crashed on the ice of Lake Keuka, near Hammondsport. Baldwin escaped injury, but the *Red Wing* and her engine were demolished. The photograph above shows the crash, an important event because it led to the development of the aileron.

The *White Wing* in 1908. Casey Baldwin flew this aircraft for the first time on May 18th. Baldwin was chiefly responsible for the design of the *White Wing*, the first aircraft to have ailerons and the second aeroplane constructed by the associates. The ailerons are the triangular strips of silk at the tips of the wings. Dr. Bell suggested that method of lateral control: hinged-surface, wing-tipped flaps. Modern aircraft now have ailerons on the trailing edge of the wing.

The A.E.A. team is shown with another aviation pioneer, Augustus Post (right), a famous balloonist who visited the experimenters at Hammondsport, New York, in May 1908. Post won the national ballooon race that year and in 1909 established a free balloon duration record of forty-eight hours and twenty-six minutes, ascending to a height of 24,200 feet. McCurdy was using crutches following a motorcycle accident.

Glenn Curtiss behind J.A. Douglas McCurdy in the *Silver Dart* when it was fitted as a two-seater. Hammondsport, New York.

Skaters bring the *Silver Dart* into take-off position at Baddeck on February 24, 1909. Douglas McCurdy sits at the controls. A few minutes after this picture was taken, he made a flight of four and a half miles. Glenn Curtiss sits perched on the wing. Brought from Hammondsport, the *Silver Dart* was reassembled. The mechanics put it together but did not immediately install the Curtiss engine. Instead, the engine was mounted in Dr. Bell's "Cygnet II" kite, an aerodrome-like device with a canard elevator system at the front and tetrahedral-celled wings. On February 20 and 21, 1909, several unsuccessful attempts were made to fly the "Cygnet II." Then on February 22, 1909, the Curtiss engine was remounted in the *Silver Dart* for a successful first flight off the ice of Baddeck

McCurdy is shown above at the controls of the *Silver Dart* in Hammondsport.

Bay on the following day. Note below the tricycle landing gear and the large push-type propeller. The *Silver Dart* weighed about 800 pounds.

Right: The Curtiss Engine for the *Silver Dart* in position for tests on an ice-boat.

The *Silver Dart* in flight over Baddeck Bay on March 23, 1909. The photograph below is similar to the design used on the commemorative five-cent stamp issued by the Canadian Post Office Department on February 23, 1959 to celebrate the Fiftieth Anniversary of powered flight in Canada.

Another flight—this time the *Silver Dart* has ski-type runners instead of back wheels. Below, the *Baddeck No. 2* in the air March 11, 1910. *Baddeck No. 1* was the first machine constructed by the Canadian Aerodrome Company and the first successful aircraft built in Canada. The wings of the *Baddeck No. 1* were covered with a type of cloth used for light sails on racing boats instead of the rubberized silk used on the *Silver Dart*.

In 1956, LAC Lionel McCaffrey built a non-flying replica of the *Silver Dart* called *Silver Dart II.* It stole the show at the 1957 Air Force Day. The Air Force permitted Lionel and other volunteers to modify the aircraft to be flown in Baddeck for the 1959 anniversary celebrations of the 1909 flight. It was tested with a few flights in Trenton, Ontario; then it was dismantled, crated and shipped to Cape Breton. Monday, February 22, 1959, the shed walls where she was reassembled were pulled away. As in 1909, a horse-drawn sleigh towed her into place on the ice. A military honour guard was inspected. A pipe band escorted her to the flight line.

A reporter wrote: "The Wing Commander Paul A. Hartman started his run on time to the minute of the flight fifty years previous. The plane rose in the air and was under perfect control when right in front of the crowd a violent gust met the aircraft and put it into a thirty-degree, nose-up attitude. Hartman recovered and pulled out of the ensuing dive just clear of the ice. More gusts struck the aircraft and the Wing Commander overcame them for a moment or two before finally coming to a hard landing which would have been much worse had it not been for his superior flying skill. There was a breath-taking moment for the crowd and an audible sigh of relief went up as he stepped out unhurt."

J.A. Douglas McCurdy in the RCAF replica of the *Silver Dart* with builder LAC Lionel McCaffrey in 1956.

Below: Wreck of the *Silver Dart II*. Note the ambulance and the broken wheel. The plane was repaired and has been on display in Ottawa at the Canadian Aviation and Space Museum since Opening Day, October 25, 1960.

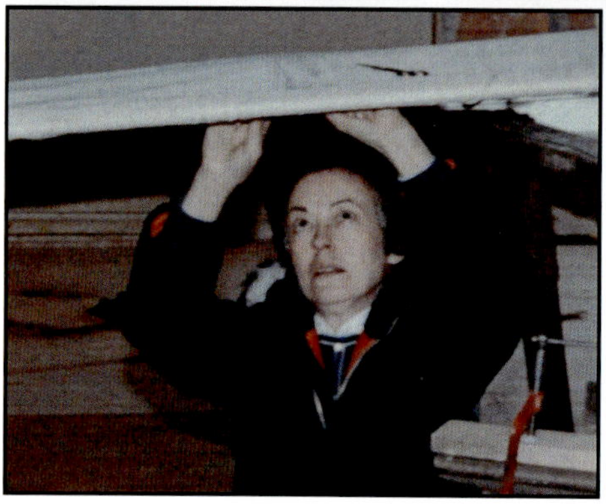

In 1984, the Cape Breton Development Corporation (DEVCO) agreed to fund a reconstruction of the *Silver Dart* for the 75th Anniversary of the first flight in Canada. Using the original A.E.A. plans, Gordon MacRae (upper left) and the team of Clarence Roberts, Kitchener MacDonald, John Johnson, Cliff Donovan, along with Mary Roberts as seamstress, built the airplane in the Masonic Hall, Baddeck. The engine was wood and the plane never flew. It went first to England and ended up at the Reynolds-Alberta Museum in Wetaskiwin, Alberta. Then the same team built a second replica of the *Silver Dart*. This one was lifted by a crane and towed by a truck, simulating flying for the movie called *The Sound and the Silence*.

The photo below is that second version of the *Silver Dart*. It is now on display at the Atlantic Canada Aviation Museum near the Stanfield International Airport outside Halifax.

In February 2009, the above replica of the *Silver Dart* was flown by Canadian Astronaut Bjarni Tryggvason at Baddeck, Cape Breton, during the centennial celebrations of J.A.D. McCurdy's historic 1909 flight. She was built by a team of volunteers incorporated as the AEA2005. Of those who were there that day, standing, left to right, are Gerry Bettridge, Kathy Bettridge, Dave Gladman, Jaro Petruck, Gerald Haddon, Carol Harding, Doug Jermyn, Don Feduck and Ray Larson. Kneeling in front: Ron Zwarych, Bjarni Tryggvason and Russ Pooler. Missing is Randy Brinkman. Photos: Vaughan Merchant.

Through a hole in the wall, the *Silver Dart*, built by the AEA2005 team and flown in Baddeck in 2009, enters her permanent home at the Alexander Graham Bell National Historic Site. Interior photos: Valerie Mason.

Great-Grandson of J.A.D. McCurdy, Gerald Haddon, with his wife Amanda, at the ceremony naming the J.A. Douglas McCurdy Sydney Airport in Cape Breton Island, Nova Scotia, July 27, 2009. Photo: Peter Arsenault.

9

The Army Is Unimpressed

Dr. Bell was so certain that the Aerial Experiment Association had a major contribution to offer the world that just a month after the flight of the *Silver Dart* at Baddeck, on March 27th, 1909, he made it the subject of a very important speech given to the Canadian Club of Ottawa. Present at the address were such distinguished guests as his Excellency the Governor General, Earl Grey; the Prime Minister, Sir Wilfrid Laurier; the Honourable W.S. Fielding, Minister of Finance; the Honourable William Patterson, Minister of Customs; the Honourable George A. Foster; the Honourable Sir Frederick Borden, Minister of Militia and Defence; the Honourable Sidney A. Fisher, Minister of Agriculture; Robert L. Borden, then the leader of the Conservative Party; Sir Sanford Fleming, Judge Idington, Edward C. Grant, who introduced the first motor car to Ottawa, and Captain F.C.T. O'Hara, Deputy Minister of Trade and Commerce.

Bell was always a popular speaker. He had a fine mastery of the English language, he had a natural poise and dignity, and as might be expected from a former teacher of speech, his diction was flawless. He said:

"I do not know but that I have a little Canadian feeling in me too. I am a good American citizen and have been for thirty-five years, but my heart has still a warm spot for Scotland, my native land, and for Canada, the home of my early manhood.

"It is a rather curious thing to me to see the dispute about where the telephone was invented. I have not any doubt about it. It was I who invented the telephone, and it was invented wherever I happened to be at the time. It so happened that I resided in Salem, Massachusetts, and carried on work in Boston. Then I would come up to my father's home in Canada to spend a large portion of my time. I carried my instruments with me, but of this you may be sure, the telephone was invented in Canada.

"It was made in the United States. The first transmission of a human voice over a telegraph wire, where the speaker and the listener were miles apart, was in Canada."

With such remarks appealing strongly to the patriotic feeling of the gentlemen present, Bell went on to discuss the early history of aviation and finally spent quite some time talking about the dirigible. In view of the fact that this speech was made nearly six years prior to the outbreak of the First World War, his genius for prophecy is rather remarkable. He said:

"Dirigible balloons have features that should make us pause. It is a thought, it seems to me, for the British nation, supreme upon the waters, to consider that a balloon, such as that of Count Zeppelin, could float over London, and all the British fleet could not prevent it. Of course, we do not know what Great Britain is doing. But the success of such a machine as that means more to Great Britain than to any other major power, because when these machines are used for purposes of war, sea power becomes secondary to air power. The nation that controls the air will ultimately be the foremost nation of the world. So, the success of the dirigible balloon, even though to my belief it is on a wrong basis, being lighter than air, is of the greatest importance to mankind."

Somewhere in his speech too, Dr. Bell defended his persistent use of the word "aerodrome:"

"People generally speak of flying machines as aeroplanes, but they are not aeroplanes, for they have not a flat surface. Our aerodromes in Baddeck are curved in a particular way, and there is not an aeroplane in the whole machine. I prefer the word suggested by Professor Langley, Secretary of the Smithsonian Institution, namely, 'aerodrome,' which is derived from the Greek and conveys the idea of traversing the air."

And with a stubborn note which was not altogether in keeping with his genial character, he said:

"I will continue to use the word 'aerodrome' to designate the flying machine itself until such time as some other meaning can be authoritatively defined."

But Bell had come to Ottawa for the primary purpose of acquainting the nation's great with the developments of flight over Baddeck Bay. First he gave the early history of his efforts at kite flying:

"I had been interested in kite flying, not the kites that little boys fly with strings, but big structures that would lift a man into the air. I was interested in this because of its bearing on the subject of aeronautics. I wished to carry a man into the air, then add a propeller and see what it would do. My kite differed from other kites in one important respect. Other kites, in a gust of wind, will dance about. But my kites for some reason are perfectly stable in the air even under circumstances of gusty wind.

"In aerial machines, what we want above all things is automatic stability, and this stability is possessed by these tetrahedral kites. I wanted to see what could be done with an engine to propel one of these with a man aboard. The first thing was to find a suitable man. I found I had not the necessary technical knowledge. I did not feel competent in putting up a man in one of these structures without an opinion from some competent engineer. So I associated myself with two young Canadian engineers, just graduated from the University of Toronto, to give me the necessary technical knowledge. One of these young men, Douglas McCurdy, of Baddeck, Nova Scotia, is a son of A.W. McCurdy, now of British Columbia, and a grandson of the Honourable David McCurdy, formerly a legislative councillor of Nova Scotia. He is a young engineer, full of enthusiasm, brave and fearless. He is the one who has been making the recent flights. He came to me as assistant engineer to help me design the engineering structure.

"The other is a young man of preeminent ability, F.W. Baldwin, a grandson of the Honourable Robert Baldwin, one of the founders of your country. He is a young man of twenty-nine, and one who will be a great acquisition wherever he goes.

"But there was still another thing to be done," continued the great inventor, warming rapidly to his speech. "We did not know about motors, so we tried to find the best man to help us in that respect. I brought to Baddeck Glenn H. Curtiss, of Hammondsport, N.Y., who had established there a business that was perfectly astonishing. He is now recognized all over America as our foremost motor expert.

"I was a little concerned that all the people associated with me in Baddeck and the workmen in my laboratory, except Mr. Curtiss, were Canadians. I welcomed, therefore, the late Lieutenant Thomas Selfridge of the United States War Department, one who belonged to my adopted country and had made a specialty of flying machines. He was bright enough to know that the time was coming when the United States Army would need such things, and he knew that when that time came, a young officer that knew about them would be of great benefit to his country. So he began to make himself an expert, and wanted to see what we were doing at Baddeck, in the interest of the United States War Department.

"So, here we were, living in my house: myself, an elderly man, surrounded by brilliant young men, each an expert in his own line. We became very friendly. My wife became very much attached to them all. She suggested that we form an association and agreed to provide the necessary capital for the experiments, which finally amounted to $35,000.00. With the providing of that I had nothing to do. I came in as one of these young men on equal terms. We had similar ideas. They wanted to help me and I wanted to help them. I wanted to give them a start in life and develop their individuality. So, we agreed to work together purely in the interest of the art. As Lieutenant Selfridge said: 'All we want is to get into the air.' All our machines are joint productions, but each one has a machine built on plans approved by him. We have five finished and call them 'dromes.' In fact, we speak of 'droming' from place to place. I do not know whether the word will take or not."

Now Dr. Bell came to the crux of his speech, that of urging that the Canadian government do something to recognize the new science which was so rapidly beginning to stir the heavens:

"Some of our fledglings have proved their ability to fly; we are going to let them fly and look out for themselves. Personally, I do not care about commercial matters. I will go on with my experiments in tetrahedral structures. Mr. Curtiss intends to continue the manufacturing of aerodromes at Hammondsport. But with Mr. McCurdy and Mr. Baldwin, the case is a little different. They say they are Canadians, and they want to go into the practical manufacture of these machines. But they say: 'Cannot we do anything for the Canadian government?' I told them I did not know, but I would be glad to help. I said: 'I am going up to Ottawa to talk to the Canadian Club, and I believe it probable that I shall have a seat at the same table with His Excellency and the Honourable Mr. Fielding and some other members of the Cabinet. Perhaps something will turn up. We cannot tell.'

"Though I cannot claim to be a Canadian, except that I have a warm spot in my heart for Canada, I do want Canada to have the benefit of these Canadian boys."

The inventor sat down to a tremendous round of applause, and some of the great men present pretended that they were in favour of immediate action. Said the Honourable W.S. Fielding:

"I have already called the attention of His Excellency to the great achievements of Mr. McCurdy and Mr. Baldwin, and am glad to say that he has taken a very definite interest in their efforts, and has reported to the Imperial authorities the results of their experiments. Our own war department has not yet indulged in the luxury of an air ship. I do not know what may happen. Of course we have made our pious resolves against expenditure, but after the address that we have heard today, I fear that Sir Frederick Borden will have dangerous intentions upon the finance department."

Earl Grey, the Governor-General, also had a few words of agreement to utter, but he spoke with the caution and diplomacy befitting his station. In a remarkably short time, however, he was to become one of the greatest flight enthusiasts of the day. And one of McCurdy's fondest memories is a special trip which Earl Grey made to Baddeck Bay for the express purpose of witnessing the flights in progress there. He came in his own private railway car, accompanied by Lord Lascelles, and he was so intrigued by what he saw that he stayed a whole week.

But there was as sharp a division of opinion in Ottawa over the aeroplane as there was everywhere else in those days. While many a solid citizen had been moved to a profound admiration for the marvel of the new machine, and were courageous enough to predict a brave future for it, there were many more who thought it the height of worldly folly.

Birdmen were not the only great men of that day, and in the summer of 1909 a sky pilot of a different order, one of America's best-known evangelists, made religious history by proclaiming to a great revival crowd that flying must, by its very nature, be evil.

"If God had wanted us to fly," he declared, "would He not have given us wings?"

There was nothing in McCurdy's Presbyterianism which troubled his conscience about his chosen profession. And while Ottawa moved in its mysterious way, and ponderously formulated plans for a flying demonstration before its Army officials, McCurdy and Baldwin were still experimenting in Baddeck. Apparently they were attracting an increasing number of press men and observers of other kinds, for there was scarcely a newspaper in America that did not contain some account of their endeavours that year. But the reaction which was of paramount importance to Baldwin and McCurdy was that of the Canadian press. Here is an account published in the *Moose Jaw Times*, June 4th of that year, and while it is perhaps more optimistic than many other dispatches published at the same time, it furnishes good proof that Dr. Bell's words before the Canadian Club had been read carefully by the country at large:

"History is being made in the quiet little watering place of Baddeck, on the shores of the beautiful Bras d'Or Lakes. Here, at the laboratories of Dr. Alexander Graham Bell, inventor of the telephone and founder of the Aerial Experiment Association, an army of experts is daily engaged with head and hand thinking and working out the problems of the best way to navigate the air. Even as I write these lines a distant purring sound comes to my ears; nearer and nearer it draws, the noise of the rapidly generating engine and the whirring of the propeller become louder and louder, until finally a window is flung open and there, sailing gracefully through the air at an altitude of about thirty feet, is the famous aerodrome *Silver Dart*, making one of her cross-country flights. At a distance of three quarters of a mile, the drome becomes lost to sight, but in less time than it takes to write these words, the machine has rounded the outer mark of the official course and is on her way back to the starting point, the aviator bringing her to earth with the graceful and unhesitating motion of a bird. Hourly flights of four miles or more are but incidents, and the aviators, Messrs. J.A.D. McCurdy and F.W. Baldwin, are as much at home seated on the machine flying through the air as is the ordinary individual sitting at the parlor fireside.

"In October, 1907, Dr. Bell organized an Association to be known as the Aerial Experiment Association. The association consisted of five members, and had as its object the building and improvement of heavier than air machines. Experiments were first made with a large tetrahedral kite at Dr. Bell's summer home in Nova Scotia. The late Lieutenant Selfridge went up in this man-lifting kite, and it was hoped to get data as to the lift and what is technically called drift or resistance, with the view to installing a motor and propellers to convert the kite into a free-flying machine. The flight was entirely satisfactory, but unfortunately the kite was wrecked by being pulled through the water after it had come down.

"The association then moved its headquarters to the engine works of Mr. G.H. Curtiss, who was executive head of the association. Gliding experiments were commenced, but it was not until the arrival of their fourth aerodrome, the *Silver Dart*, that substantial successes were obtained. The *Silver Dart* is a double decked aerodrome, 49 feet wide, with planes six foot broad, fore and aft. Centred on it is a 50 horsepower eight cylinder motor, built by the G.H. Curtiss Company of Hammondsport, New York. This motor drives a two blade wooden propeller, placed at the rear of the airship. The elevation is controlled by a horizontal rudder, broadly made, and extending out in front, by the tilting of which the aviator can direct his force up or down. At the rear is placed a perpendicular rudder, by which the drome is steered, and which operates like that in an ordinary boat. The planes, with the exception of two small triangular wings at the extreme tips, are of rigid construction, the sections being covered by vulcanized silk. The forward thrust of the propeller, when the drome is stationary and the engine is going at full speed, is 300 pounds.

"March 10th witnessed the sensational flight by the *Silver Dart* when the machine outdistanced all her former records. Having been fitted with runners, the machine was trundled out on the ice, where, after a preliminary examination of the motor and machinery, Mr. J.A. D. McCurdy, owner of the machine, climbed through the network of wires into his tiny seat in front of the engine. The body of the machine was held in check by a half a dozen sturdy onlookers while the engine was set in motion. With a 'Let her go!' from the operator, the attendants released their hold, and away went the machine over the ice like an arrow sprung from a bow. After skipping along the ice for about 200 yards, the drome rose to a graceful angle to a height of about twenty feet and shot away for the town of Baddeck, some three miles distant. Passing over the channel between the town and Kidston's Island, the aviator headed the machine for the ten mile turning buoy. Rounding the mark in a wide, sweeping circle, the drome started on her return course. On the way in she had to pass over a thickly wooded patch of land, but was sufficiently high to avoid coming into contact with the tree tops. The finishing point was rather close inshore, and Mr. McCurdy, to avoid the chance of landing on shore, shut off the power, and the machine, bereft of motion, glided gracefully to the ice. This was the record flight of the *Silver Dart*, the drome having covered a distance of twenty miles, less 200 yards, in twenty-four minutes.

"Needless to say, this little group of men who are working so earnestly for Canada's success in the science of human flight are enthusiastic in their hope for the future.

"'I feel confident that this machine should continue flying for an indefinite period,' the aviator says. 'These flights we are making are only experiments, and occasionally during a flight, when we have gone scarcely more than 100 yards, some little piece of mechanism gets out of adjustment, and we have to alight to fix it. Then again, the machine will run along, as you have seen, for twenty miles without a hitch. The aerodrome is here to stay, and after it is somewhat more perfected, I look to see it employed by the big powers of the world during times of war to carry dispatches, do scouting work, etc.'"

Perhaps of more scientific interest to the world at large was the reaction of the *Scientific American*, which that same month wrote as follows:

"Mr. J.A.D. McCurdy, the secretary of the Aerial Experiment Association (which has lately disbanded), was in New York recently. He stated that during the past winter he had flown more than 1000 miles, having made over 100 trips over a nine mile course, carefully measured and staked out upon the ice of Lake Bras d'Or. The machine used was the *Silver Dart*. Messrs. McCurdy and Baldwin, of the A.E.A., have formed a company to manufacture similar machines in Canada. They have already nearly completed two aeroplanes similar to the *Silver Dart*. [These were the *Baddeck No. 1* and the *Baddeck No. 2*.] Among other things, Mr. McCurdy rather upset the theory of wind gusts affecting an aeroplane in flight perceptibly, by stating that in his flights in Canada he passed over promontories and near hills without having his machine affected in the least. At one time, when flying at Hammondsport, the end of one wing struck a branch of a tree as big as a man's finger, and broke it off, without having the equilibrium disturbed. In view of the above, and also of the practical demonstrations of cross-county flight already made by Farman and Bleriot, it does not seem probable that slight aerial disturbances will stop a machine from following the course of the Hudson River from New York to Albany, and winning the Fulton Flight in case the motor does not give out."

As revealed in this article taken from the pages of the *Scientific American*, the Aerial Experiment Association disbanded on March 31st, 1909. Casey Baldwin and McCurdy were now free to plot a different course. Neither entertained any thought of leaving Baddeck and Bell, however. The result was that the spring of that year saw the formation of Canada's first aircraft company. This was the Canadian Aerodrome Company, which Bell financed and which he had duly registered by a special act of the Nova Scotia legislature. In the fond hope that the Canadian government would soon add aircraft to its defence machine, plans were made for the manufacturing of craft somewhat similar to the *Silver Dart* in design and capability. That next spring and summer saw the manufacture of two of these, the *Baddeck No. 1*, and the *Baddeck No. 2*. And the young flyers were testing all three craft very thoroughly on Bentinck's farm when the word came back from Ottawa that arrangements had been made for them to demonstrate the art of flying at the Army's great camp at Petawawa.

Baldwin and McCurdy crated the *Baddeck No.1* and the *Silver Dart* and shipped them to Ottawa. When they arrived there, they were surprised to learn that a hangar had already been constructed, the first of its kind on Canadian soil. But it was quite evident that the optimism which had apparently been responsible for the hangar was not to be found in all of the officers present.

On the morning of Monday, August 2nd, the *Silver Dart* was duly paraded before the officers present and made five flights. She flew about fifty miles per hour at an altitude of fifty feet, demonstrated her turns and her inherent stability, and flew distances of about a mile in length.

All flights were thoroughly successful and without incident of any kind, except the fifth, and here McCurdy suffered the only serious injury of his long career. The accident was due to McCurdy's unfamiliarity with the sandy soil conditions over which he was flying. Glenn Curtiss' slight motorcycle wheels had done their job admirably on the clay or sod at Baddeck and Bentinck's farm. But on the landing after the fifth flight, the wheels sliced into the Petawawa sand so deeply that the *Silver Dart* went over on her nose and was demolished beyond any hope of reconstruction. McCurdy came out of the wreck with a broken nose.

Thus ended the life of the *Silver Dart*, sacrificed in the name of King and Country. Today, a monument marks the spot of her death and proclaims it the birthplace of Canadian military aviation. Officials who watched the mishap, however, did not think of her demise in any such patriotic manner. The fifth flight, so they decided afterward, was really the official flight. And since this one had been a failure, they solemnly ruled that flying seemed to have no place in modern defence. The decision was duly recorded in triplicate and passed back to the War Office. Even in the matter of the monument many years later, the Army continued to dispute McCurdy, as though it were now a tradition to do so. The plaque on the cairn maintains that Baldwin flew the *Silver Dart* along with McCurdy. He didn't.

Baldwin had witnessed the proceedings from the ground and it was he who first sensed the reaction of officialdom and who quietly informed McCurdy that day that "the nays have it."

Then as now, the national caution was the national weakness. Our young men have always been brilliantly brave, and eager to bring dreams to life. But the men to whom they owe allegiance are invariably older, safer men, who have a more proper regard for tradition and sanity.

10

Barnstorming Days

There was a low tide now in the affairs of the men from Baddeck. They had lost out on their bid for favour from the Army, and the proud craft which had made them famous was utterly ruined. The *Silver Dart* was indeed so completely smashed that McCurdy didn't even try to salvage the stout little motor. It was taken out of the wreck and finally found its way back to the east, where it was used to power a fishing boat. A year or two later, as if to complete the humiliation, the ship whose heart she was went to the bottom one day. There she stayed with the mud filling her cylinders and the salt brine eating at her copper jacketing, until the gratitude of a more enlightened age searched her out and polished away the corrosion of the forgetful years so that the sun could have another look at her. Today the engine of the *Silver Dart* sits in an honoured nook in the National Research Council building in Ottawa, the same city that had once laughed at her tragedy.

There were a few more flights with the *Baddeck No. 1* and the *Baddeck No. 2* over the Bentinck farm that fall of 1909 but there was not much in the way of industry in the newly formed Canadian Aerodrome Company. Furthermore, Casey Baldwin had pretty well decided now to accompany the Bells on a round-the-world trip which they planned for the following spring, so the year ahead promised to be a rather lonely one for McCurdy.

However, McCurdy had one great accomplishment which was very much in his favour. He was certainly the best flyer of the five associates. In all probability he was the most skillful flyer in the world that year. And the curious world was evidently eager to see flying demonstrations. They might scoff at any suggestion that the new miracle would revolutionize the transport of the future, or that it could be a mighty comfort for the country in time of war, but they were still willing to flock by the thousands to watch any man who promised to take to the air. And they were willing to pay hard-earned money for that privilege just the same as they would to witness any other daredevil act where a man might break his neck before their eyes.

In the States, Glenn Curtiss was performing before enthusiastic crowds of ever-increasing size, and he was managing to make good money at it. Why couldn't McCurdy do the same? So it was that the year 1910 found the name of J.A.D. McCurdy on a newspaper's front page in practically every large American city east of the Mississippi. His first air show, at Valois, Quebec, was a dismal financial failure, however, and he decided then to return to the States. There he gave flying demonstrations at New York, Syracuse, Washington, Memphis, Atlanta and in many other smaller cities. The business arrangements for such aerial circuses were very simple indeed. McCurdy's procedure was to approach a leading newspaper in the city to be visited, and to get that paper to sponsor his visit as an advertising stunt for them. His brother Lucian acted as his public relations man and treasurer.

A good example of the way that a McCurdy visit was arranged for a city is given in the following dispatches from the *Norfolk Ledger-Dispatch*, a Virginia newspaper of December 21, 1910. Headlines for that date were these:

McCURDY WILL FLY OVER NORFOLK
TOMORROW, WEATHER PERMITTING

Underneath this banner the following directions are given for the benefit of the citizens:

"Signals for McCurdy's Flight

"The Norfolk office of the Weather Bureau promises good flying weather for tomorrow—light southern winds and temperature between forty and fifty degrees.

"Fifteen minutes before the flight starts a flag will fly from the Bank of Commerce building.

"This will be a signal for all the whistles in the harbour to blow. The whistling will continue until Mr. McCurdy is seen in the air. Flight will start some time between noon and four o'clock. The start will be made from Orapex Avenue on the waterfront."

Elsewhere in the newspaper a columnist elaborates as follows:

"J.A.D. McCurdy is here and will fly over Norfolk, weather permitting, between noon and four o'clock tomorrow.

"The mystery that surrounds the hangar on Red Gate Avenue has been lifted.

"The strange tent that housed an aeroplane has been holding the machine Mr. McCurdy will use tomorrow, the famous *Hudson Flyer.*

"The *Ledger-Dispatch,* unless the weather again makes the attempt impossible, will tomorrow fulfill the promise it made some weeks ago to the people of Norfolk.

"Just adjoining the Westover property at Orapex Avenue and the waterfront, McCurdy will leave the ground. Gradually rising higher and higher until he reaches a height of 2,000 or 2,500 feet, he will follow the river until he comes to the east end of Norfolk. Turning then he will fly above the city, going over the very center of the town. Then he will turn back to the starting point and if possible make a second detour.

"Some time, the most favourable between 12 o'clock and four in the afternoon, the daring aviator will fly. From the top of the Bank of Commerce Building a flag will be flown fifteen minutes before the start. When this signal floats on the wind a babel of whistles will sound. Every available boat in the harbour will add its quota and everyone should be thus apprized of the aviator's approach. The *Ledger-Dispatch* trusts that the weather man is right in his forecast for tomorrow and promises that, if he is, the people will see the flight on schedule time. McCurdy came here from Atlanta solely for this purpose."

Headlines in the same paper the next day were these:

McCURDY IN HIS AEROPLANE SAILED

HIGH ABOVE CITY OF NORFOLK TODAY

"While thousands gazed upward in wonder J.A.D. McCurdy soared high over Norfolk this afternoon in his aeroplane, thus fulfilling the pledge of the *Ledger-Dispatch* made weeks ago. McCurdy rose from the ground near the waterfront on Orapex Avenue. Higher and higher he soared, then stretched away following the course of the river around Norfolk. Circling back he landed at his point of departure ten minutes from the time that he started. The flight was made without any mishap.

"'Gee, but it was cold up there!' said McCurdy as he landed."

In general, McCurdy worked south as the cold weather crept southward. He apparently figured that $500.00 was the minimum amount which he would have to have in order to make a visit to a city worth while. Each visit meant that the plane would have to be crated to its new destination, and there reassembled and retested. He took his own mechanics with him, and a tent was pitched as headquarters on some suitable flying ground at the city's edge. He became more skillful as the weeks progressed; his plane flew to new heights and attracted bigger crowds.

In Florida it appears that McCurdy was also welcomed into high society, for the following dispatch, from Ormand, Florida, reads as follows:

"J.A.D. McCurdy, the aviator, who is making exhibition flights at Sea Breeze, Florida, made a flight to Ormand today in order to take tea with a number of Bostonians whom he had met at Palm Beach. The tide was too high for an auto and there were no afternoon trains, so McCurdy, after telephoning, took his biplane and sailed up the beach to be met by Miss Elsie Pinkham of Lynn, Miss E. Phyllis Kimball of Providence, and Dr. J. Wilson Parks and H.P. Farrington of Boston. Other New England visitors at tea were Miss J.B. Pinkham and Mrs. Laura Hunt of Lynn, Mr. and Mrs. J.L. Damen of Boston and Mr. and Mrs. H.A. Kimball of Providence.

"After tea McCurdy sailed back to Sea Breeze, with an admiring throng looking on."

It was while McCurdy was sunning himself in Palm Beach that he first began to look longingly across the blue waters of the Caribbean toward the tropical shores of Cuba. No one had ever made an ocean flight of such a length, but McCurdy firmly believed himself competent to accomplish that now. He knew no Spanish, and had no connections in Cuba, but since Havana had a newspaper the same as large American cities, he saw no reason why he could not find a sponsor over there.

The finding of sponsors was much easier than he thought. The *Havana Post* offered to pay him $5,000 if he completed a non-stop flight from Key West, and the city of Havana offered an additional $3,000.

McCurdy was jubilant and made his plans for the flight at once. He found an equally important ally in the U.S. Navy, and the destroyers *Paulding* and *Drayton* were instructed to give all aid possible to him in mapping out a course. The unreserved support that the U.S. Navy was willing to extend to McCurdy is indicated in the report from the *Key West Citizen* dated January 23rd, 1911:

"The U.S. destroyers *Paulding* and *Drayton* returned last night from a quick trip to Havana with Aviator J.A.D. McCurdy and party.

"The trip was made to acquaint the aviator with the course and to do some experimenting on the manner of marking the course by means of smoke from the destroyers, also to give Mr. McCurdy an opportunity to select a landing place in Havana.

"On Saturday morning at 10 o'clock the *Paulding*, commanded by Captain Stirling, left the oil ship *Arthusea* and steamed out of the harbor, closely followed by the *Drayton*, commanded by Captain Dinger.

"After reaching a point about ten miles south of Sand Key, the *Drayton* was stopped, the *Paulding* continuing at a speed of about 20 knots, which was later increased to 24 knots. When the hull of the *Drayton* began to fade away she was caused to make smoke, which enabled her location to be distinguished for a much longer time. Her hull was lost sight of by the *Paulding* at 10 miles, but the smoke could be seen for about 17 miles.

"Morro Castle was sighted by the *Paulding* at 1:45 and she passed into the harbor at 2:40. The *Drayton*, which had again followed, arrived a short time later. During the trip Mr. McCurdy made numerous observations and consulted from time to time with Captain Stirling regarding plans for the big event.

"Upon arriving in Havana Mr. McCurdy and his managers conferred with the management of the *Havana Post*, under whose auspices the flight will be made, and arranged other details. The aviator also visited Camp Columbia, where he hopes to land, and also several other points where it might be convenient to land under certain conditions were inspected.

"On the return trip yesterday evening, plans for the rescue of the aviator and his machine in case he falls into the water were discussed. Work was started to build platforms on one of the destroyers large enough to hold the aeroplane and from which a new start can be made in case the machine is not badly damaged. The platform will be on the *Paulding*.

"The plucky aviator does not intend to let a little thing like a dip in the water stop him, if he is at all able to make a fresh start from the deck of the destroyer.

"He expects to fly at an elevation of about 1,000 feet on the trip to Havana and to make about 50 miles an hour."

A little later in the same dispatch, a reporter aboard the *Paulding* seemed quite proud of the fact that he had the privilege of having several pleasant chats with the famous young aviator from Canada, and found him to have a most fascinating personality. Said he:

"There is nothing odd or freakish about McCurdy; he is just a normal person of rather exceptional attainments, whose most distinguishing characteristic is his modesty when discussing himself.

"While not a total abstainer, Mr. McCurdy drinks sparingly of alcoholic beverages, though he smokes moderately, either pipe, cigar or cigarette. He follows no special diet, either when preparing for an unusual flight or otherwise, neither does he have any set rules for sleeping."

McCurdy apparently was well aware of the fact that he might not be able to claim the $8,000 which was awaiting him across the waters in Havana, for he planned one last flying exhibition on Trumbo Island. This time he got the Key West Chamber of Commerce to sponsor him, and the whole island was rented for the demonstration.

There seems to have been some doubt in the minds of the sponsors as to whether or not a sufficient number of Key West citizens would want to pay to get on to the island when it was very clear that they would see the most of the flight from their own back porches. Hence this unique bit of advertising in the press the day previous to the flight:

"The price of admission to the Island has been fixed at $1.00 each per adult. There are many things to be seen in connection with the flight that can be enjoyed and appreciated only by those who are within the enclosure. This price is below the usual charge made at aviation meets and is done as a special favour by Mr. McCurdy. The people should take advantage of this offer and buy a ticket as the aviator will not fly unless there are 500 paid admissions.

"The flight to Havana for the *Havana Post* Prize will be made so early in the day that but a few will have an opportunity to see it.

"Tickets are now on sale and to those who have any doubt about coming, the following letter from Mr. Maloney, President of the Chamber of Commerce, should influence them to buy a ticket and see the flights properly:

"'I have consulted Mr. Cotton and his consent has been granted. So Mr. McCurdy will treat the people of Key West on Tuesday afternoon at three o'clock to the greatest sight they have ever seen.

"'Mayor Fogarty has kindly tendered the police force to keep those off Trumbo Island who have not obtained tickets. Five hundred dollars must be raised, else I doubt very much if we will have the pleasure of seeing the flight.

"'Those who stay outside and get on the house tops will lose the best part of the exhibition. While I was in Philadelphia I saw Grahame-White make several ascensions and to me, the most interesting part was the getting off and going up of the machine and the landing. This cannot be seen from a distance, besides those who will be in the enclosure will have an opportunity of inspecting the aeroplane at close range. This alone is worth the price of admission. We should make Tuesday afternoon a holiday.

"—J.B. Maloney'"

After the exhibition at Trumbo Island, McCurdy made the last feverishly urgent preparations for the big flight to Havana. This was something he dared not lose, but even if successful he would not be the first of the five associates to win a similar prize. On May 31, 1910, Curtiss flew down from Albany to New York, following the Hudson River to win his second *Scientific American* trophy as well as a prize of $10,000 offered by the *New York World*. One incident in that flight provides a graphic illustration of what the world thought in those days of these wild young men who were so willing to risk life and limb to conquer space and the law of gravity.

In his flight from Albany to New York, Mr. Curtiss was allowed to make two stops. He planned to make the first of these at a place near Poughkeepsie, and the second at the upper end of Manhattan Island. Prior to the flight he visited these localities and at Poughkeepsie he found a large, flat, beautiful lawn. It would be ideal for a landing place, so he went to the proprietor of the institution which apparently owned the ground and asked permission to land there. The answer was cordial and very much to the point. "Certainly, Mr. Curtiss, we shall be very glad to have you land." The speaker proved to be Dr. Taylor, head of the institution. "This is the place where they all land," he said.

It was State Asylum for the Insane.

11

Havana and the Cuban Sharks

The Havana flight made McCurdy the darling of all America, and the adulation accorded him was something more than the hysteria which ordinarily rewards spectacular success. The heart of the whole continent was with him from the moment he announced his daring plan, and though McCurdy had to wait a disappointingly long time before he could actually point the nose of his plane toward Havana, the support of his fans never wavered for a moment. Indeed, the lad from Nova Scotia was credited with a courage of such gallant proportions that the press of the day bestowed upon him the greatest honour that America can offer a foreigner. It found a way to label him an American.

Said the *New York Herald* a week or two before the epoch-making flight: "McCurdy is a Canadian by birth, but an American by aviation."

His citizenship having been duly rectified, the United States Navy now took the young aviator to its great heart and put a whole flotilla of naval craft at his disposal. In all there were six torpedo-boat destroyers assigned to patrol the line of flight between Key West and Cuba that January of 1911. Briefly the plan was this: the *Paulding* with Lieutenant Commander Stirling in charge would lie just outside the Havana harbour. Stretched between the *Paulding* and the Key West coast, and spaced about ten miles apart were the *Terry*, the *Drayton*, the *Roe*, the *Arthusea*, and finally the cutter *Forward* would be stationed just off the Key West coast. All boats were equipped with wireless, and weather conditions could thus be carefully checked for the whole ninety-mile stretch of water to be traversed. Furthermore, the Navy thought that by spacing their ships across the sea in ten-mile intervals, McCurdy would never be out of their sight. Wireless would thus keep all ships fully informed of the aviator's progress, and should he be seen to tumble into the sea, there seemed to be at least a fifty-fifty chance of a ship being able to effect a rescue.

It was an elaborate plan, and the Navy's enthusiasm and good will kept everyone in high spirits.

McCurdy's brother Lucian laconically observed: "We've got a five million dollar escort." On this Havana adventure, as in all of McCurdy's other barnstorming stints, Lucian was along to act as business manager.

McCurdy had one other thing in his favour. The associates had not yet solved the problem of the *Loon's* stubborn duck feet (the "step" was not invented till 1911) but the plane which he now had was equipped with a pontoon attached to each wing tip. McCurdy had had them made for him by a Key West tinsmith and they were designed to keep his craft afloat should it have to come down. It was doubtful, however, if these pontoons could keep him afloat for very long, particularly if the water should be rough. But at least they might give a man a little more time to keep his head above water.

But now came an obstacle which neither the genius of inventors nor the power of the United States Navy could conquer. Day after day a big wind kept blowing in from the north. Such winds were most uncommon, as any tourist agency in Florida will gladly tell you, but this was a wind which promised to go on forever, and what should have been a smooth, gentle stretch of blue off Key West was a heaving, angry sea which seemed to have no intention of ever being pacified. For seven days the norther blew. The wind was of sufficient force to keep even the sponge men, the turtle schooners, and various other exotic types of fishermen tied up at port.

The citizens of Key West had seen McCurdy flying a sufficient number of times now to have some idea of the part which weather played in flight and they understood the waiting. Across the water in Havana, however, there was no such patience. Every morning the people woke up excitedly thinking that the birdman from America must surely be on his way. And when each day's sky was as uneventful as yesterday's, the public began to attack McCurdy's name with a rather astounding brutality. The aviator was openly proclaimed to be a coward and a greedy faker. What he hoped to gain by deceiving them thus, they were not too sure. Perhaps it was just to bring ridicule upon their city.

McCurdy made no reply. In fact he may have been too busy at the time to have known the abuse which was being heaped upon him ninety miles away. He did, however, arrange to have a second plane awaiting him at the flying grounds in Cuba. In case his own craft did not survive the ordeal, and he did, Havana would at least not be deprived of its aerial circus. But one day, after the Navy had tried to tell the Cubans that the new flying machine could not be risked in high winds, a Havana weather bureau retorted that they had been experiencing a quiet sea and a five-knot breeze all week. The Navy promptly investigated the Latin claim and sent a destroyer over to feel out the Cuban coastal waters. The destroyer came back with the news that it had run into a near-gale and that the whole harbour was ringed with white water. Havana was thereupon heartily damned, and no one paid any more attention to her cries of outrage.

McCurdy's most reliable expert on the weather was his younger brother Lucian, whose intimate knowledge of flying and calm good judgment were now tremendously important to the cause. And in the plans which the Navy had worked out so carefully, Lucian was to stay aboard the *Paulding* off Havana. Here he was to judge whether or not the weather was suitable for an attempt and to wireless his decision back to his brother waiting at Key West.

On Friday night the weather bureau reported a probable let up and change of wind, but the report got through to McCurdy's party when it was almost midnight. The lateness of the hour, however, did not deter the Navy from immediately making plans.

A rare good party of naval officers and civilians was in progress at one of the local pubs that night, and when the word came through, there was a last toast and a round of applause for McCurdy. The ships put out to sea, finding their way out of the difficult harbour by searchlights.

McCurdy had been so long delayed that he was pathetically anxious to fly, but the decision was to be his brother's rather than his own.

Dawn brought only a slight slackening off of the wind. The sea was down somewhat—enough for the whitecaps to disappear. And the harbour was much calmer than it had been for many a day. But out on the ocean, and on the Havana side in particular the wind, which was supposed to subside entirely, suddenly began to blow quite briskly again. Aboard the *Paulding*, Captain Stirling was loudly impatient. He looked at Lucian as much as if to blame him for the sudden change in the weather. "Well?" the Captain said.

McCurdy's brother looked out across the waves, turned his face into the wind, and said: "I don't know."

From the wireless room of the *Paulding* a messenger came with a wireless dispatch from Key West. It stated simply that McCurdy was chafing at the bit in Key West and wanted to fly now.

The other McCurdy still had his face swung into the wind. "Wait an hour," was his answer.

Captain Stirling made no effort at all to make Lucian change his mind. He knew very well the dangers involved, even when the conditions were ideal. He was none too sure about the success of this wild idea and his chief hope was that he could save the man when the emergency happened.

The next wireless message from Key West was more impatient still. "The wind is down here," it said. "Shall I fly now?"

Captain Stirling watched Lucian read it. Lucian seemed to be a long time making up his mind. "Well?" snapped Captain Stirling. "You're the doctor!" Lucian had a painful moment of indecision. "What do you think?" he asked.

"I can get your brother in twenty minutes if he falls within sight of one of our boats."

"How about the plane ?"

"I don't guarantee anything."

Lucian scratched his head and finally wrote out an answer for the wireless operator to send back. It was: "Wait half an hour."

Within a few minutes the messenger came back with a third wireless from Key West. "Conditions perfect here," it said. "Shall I fly?"

Lucian took another hard look at the rolling water and let the wind push into his nose again.

"Well?" It was Captain Stirling again.

Lucian turned to leave the deck. "You tell him," he said trying hard to laugh, "to take his bath today at the Jefferson Hotel in Key West."

Captain Stirling did not argue. There was no flight, and the people of Havana were shriller than ever in their denunciation of the braggart who had promised that he would fly to their country.

There were, however, at least a few Cubans who gave invaluable aid at a time when it was needed most. These were meteorologists from a school of Jesuits near Havana. McCurdy and the Navy officers were soon to discover that the absolute faith which the fishermen had in the prognostications of these priests was not from religious awe but because the Jesuits' prophecies were firmly grounded on good science.They had an amazing record for accuracy.

On Saturday, January 28th, the Jesuits volunteered the information that a sort of cyclone had been sweeping the straits and that it would be followed soon by another cyclone of a similar nature. But in between the two storms there would most certainly be a brief period of calm. This, they declared, might be expected to come late Sunday, and it could not last for more than about twenty-four hours. All of which meant that if McCurdy was to make his attempt, he must do so in this brief space of quiet or not at all.

So McCurdy got the word, considered it briefly, and then made last-minute preparations for the big event. The torpedo-boat destroyers assigned to him got the word too and immediately made plans for taking their stations along the line of flight. At two o'clock on Monday morning, the destroyers left Key West and stretched themselves out to Havana. Sure enough, the next dawn was comparatively quiet, the wind had gone, and the sea was only vaguely troubled. The day came up bright and almost cloudless, and a new warmth hung over it. Men aboard the ships amused themselves watching schools of playful porpoises and flying fish. The fishing boats untied from the wharf, too, and put out to sea by the score, but more in anticipation of the famous adventurer who was to pass overhead than to resume their way of life.

The news spread through the whole of the country within the matter of an hour or so, and almost every able-bodied citizen in Key West found his way to Trumbo Island to see McCurdy off.

On the other side of the water, the Cubans began firing cannon out over the sea as a signal that at long last the great event was about to take place. And the Havanans tumbled out of their homes by the tens of thousands to the waterfront.

McCurdy lifted at 7:02 Key West time, 8:02 Havana time, and after circling Trumbo Island and the cheering crowds twice, he pointed the nose of his plane toward the first of the Naval escort ships. Weather conditions were perfect, with only the faintest of wind, and the sea showing no sign at all of changing its placid mood. Of the beauty of that sea, McCurdy said later:

"Never have I experienced so wonderful a sensation as when I rose a thousand feet and felt myself lifted out over the sea. Ahead of me was a wondrous mirage; the ocean, instead of a horizontal plane below, had the appearance now of a huge vertical picture on which the funnel tops of the destroyers appeared like black spots. Soon I became accustomed to this, and the sensation was glorious as I rushed forward at fifty miles an hour toward this wonderful panorama, which then appeared to recede from me. It seemed no time before Morro Castle and Havana were in plain view, and I could almost see the waiting crowds.

"Then something went wrong with the engine. I heard it falter, and I saw all my oil running out. There was nothing to do but descend. This I accomplished with perfect ease. The accident only increases my earnest desire to effect the conquest of the Florida Straits, and I have the fullest confidence that I can do it. I do not know when I shall be able to make another flight, but under such conditions as prevailed today I am sure of success."

The defect in the engine, as it was discovered later that day, was a burned bearing on the crankshaft. McCurdy stopped his motor at an altitude of about 800 feet, and glided down very easily and gracefully to alight on the water. There was no damage either to himself or his machine. The homemade pontoons did their job perfectly.

"I didn't even get wet feet," he said.

Back at Trumbo Island, the crowd dispersed very quickly as soon as the aviator had disappeared from view and flocked about the wireless station to hear the news. When the message came that McCurdy had passed the third station ship, a tremendous cheer went up. But as the moments flew by and no further reports came back, the crowd became very anxious and began milling about the cable office to wait for reports from Havana. And when at last the news came back that their favourite aviator had fallen into the sea, a groan went up. However, when a second dispatch informed them that McCurdy had been forced to alight on the water because of a shortage of oil and that neither he nor his machine had been injured, the cheering was almost as great as that which had sent him into the skies an hour or so earlier that morning.

Aboard the *Paulding*, the scene was somewhat different. Captain Stirling had already been informed that McCurdy had passed the *Terry* and all eyes were eagerly searching the skies in the direction where he must appear at any moment. Suddenly, a seaman spotted him without the aid of any field glasses.

"There he is!" he said excitedly.

There he was indeed—two horizontal streaks of black against the blue of the sky. He had only one more boat to pass, their own, and after that he would surely make Havana. Every man on board began to cheer and to take off his hat as McCurdy rapidly approached.

Then there was a moment of ominous silence and somebody cried: “My God! He’s fallen!”

It was true. The birdlike black speck in the sky was there no more. In a matter of seconds the screws of the *Paulding* were churning the water behind it and within a few minutes it had reached its top speed of twenty-nine knots. On the Key West side of her, the *Terry* was also steaming to the rescue and the two destroyers reached McCurdy at the same time. A boat from the *Terry* put out, and there was McCurdy in perfect control of himself, calmly standing on one of the pontoons.

“That was tough luck!” he said. “Why, I could see Havana!”

That was quite true. The coast was only a mile and a half away, Havana but two miles. And though he had not quite finished his flight, he had set a new world’s record by flying eighty miles over open water, several times the distance of the Channel flights.

McCurdy did not lament his rotten luck for long. He was too busy directing the rescue of his beloved plane. He had set it down on the water so gently that it had suffered in no way at all, but the eager men of the *Terry*, unacquainted as they were with the fragility of aircraft, damaged it rather badly when hauling it up on their ship. Their reason for haste was rather forgivable, however, because they made the dramatic discovery that they were not the only ones who had come to the scene of the accident. Swimming around in anticipation, their snouts almost breaking the water, were three fourteen-foot tiger sharks.

As the *Paulding* sailed into Havana harbour, McCurdy discovered that he was still very much a popular hero in spite of his failure. President Gomez came out to meet the *Paulding* in the Cuban flagship *Hatuey*, and when he landed, the crowd which had damned him so petulantly but a day before, now tried to smother him with affection.

“We will give these people a real show anyhow,” McCurdy decided, and proceeded to go to the drill grounds at Camp

Columbia where without even waiting to change his clothes, he gave a marvellous exhibition of flying skill, in the other plane he had waiting for him as a reserve, "just in case."

It was the first exhibition of flying that Cuba had ever seen and the whole country was in a fever of excitement. Loudest of all in the praise for McCurdy was Cuba's President Gomez. The exhibition over Camp Columbia had scarcely concluded when Gomez announced that he would start plans at once to form a military aviation school. And to honour McCurdy for his part in inspiring him to such a gallant decision, he announced that there would be a state banquet given for the aviator at which time the government would present him with a cheque for $10,000. The *Havana Post* immediately decided that it would be good policy to play "follow the leader," and proudly announced that in view of the fact that McCurdy's flight had not ended until he had come over the territorial water of Cuba, their promised cheque of $5,000 would go to McCurdy. Finally, the City of Havana said that the $3,000 they had offered would also be paid in full.

For a nation which had so roundly cursed him but a day or two before, it was a most heartening change of mood, and there was apparently even more to come. McCurdy and his associates made several exhibition flights in Havana and elsewhere in Cuba that next week, and each meeting saw unprecedented crowds and new outbursts of Latin adulation. Everywhere McCurdy went, he was besieged by a mob of hero-worshippers, and by countless influential citizens begging to bestow some favour upon him. The first day or two of this new glory was very flattering indeed, but it was not long before McCurdy and his brother Lucian were looking around somewhat breathlessly for a way to achieve a little privacy. And it was one night in a very exclusive Havana club that they met Rodriguez.

They had no reason to know the man's name that night, and indeed would probably not have noticed him had it not been for the two delectable Cuban beauties who clung very tenaciously to him, one on each arm. Rodriguez was a rather unimpressive-appearing man, short, thick and dark. That he was a man of some importance, however, was quite evident from the class of women who were now seeking to possess him, and from the fact that he did not seem to be interested in the slightest. Suddenly, as the McCurdys watched the little performance of the short man trying rather vainly to get rid of his two female companions, they saw him take a hundred dollar bill from his pocket, tear it in half, hand one section to each woman, and then shoo them out the door.

If Rodriguez knew who the Canadians were, he gave no sign that night. He probably didn't even notice them. But next day, as McCurdy was tuning up the motor of his Curtiss biplane in preparation for some more exhibition flights over Havana, he felt a tap on his shoulder and looking up was surprised to see the same man who had been so reluctant in the nightclub the night before. It was then that the man gave his name. "Would you make one flight for me?" he asked.

McCurdy confessed later that he was so busy testing his motor, that he gave the question little thought, even though he was intrigued by the coincidence.

"So I told him, sure," McCurdy said, telling the story in later years. "Why not? I was going to make the flight anyhow. It might as well be for him as for anybody else. What I didn't realize was that he wanted to make a publicity stunt out of it for his own benefit. He was one of the biggest cigar makers in all Cuba, and when I took off, it was announced to the crowd that this particular flight was being made through the courtesy of 'Romeo and Juliet Cigars,' these being the cigars that Rodriguez made."

When McCurdy came down, the same little man was waiting for him and beaming broadly he placed a cheque in his hand. It was for a thousand dollars. McCurdy tucked it into his wallet and later that afternoon when he returned to his hotel, he took it to the clerk. "Do you think this cheque is any good ?" he asked.

"I'll cash it for you right now, if you want," the clerk replied. "Rodriguez is one of the wealthiest men in Cuba."

Upstairs in his hotel room another surprise was waiting. Inside the door was a three-foot stack of boxed cigars of all shapes and sizes. There was also a very cordial invitation for McCurdy and his party to visit the Romeo and Juliet factory next day, an invitation which they accepted. They were shown through the entire factory, and McCurdy was feted as if he were a visiting dictator. Of that visit, one thing he recalled vividly. This was the practice which Rodriguez had instituted in his factory of having someone read to his workers as they rolled cigars hour after hour.

"You couldn't help but like a man who would do a thing like that," McCurdy said.

There was another reason why McCurdy always remembered Rodriguez with more fondness than he did some of the other great Cubans he met that winter of 1911. Rodriguez's cheque for $1,000 was good. The cheque for $5,000 from the *Havana Post* was never received. A photograph of it, duly signed by the publisher, was reproduced in *El Figaro*, a few days later, but where it went after that, McCurdy never knew. Nor did he get the $3,000 which the city council had so grandly proclaimed for him.

The state reception, however, did materialize. All the great dignitaries of the nation from the President on down jammed into the opera house, and the ovation given McCurdy was the most glorious he had ever experienced. Finally, when the speeches and the bands had worked the great assembly into near hysteria, came the high point of the evening and McCurdy was presented with the envelope containing the President's gracious prize. The envelope in itself was a work of art, being about eight by ten inches in size, covered with elaborate artistry on the front and with the President's seal affixed on the back. The envelope was indeed such a breathtaking work of art that McCurdy did not open it until he reached the privacy of his hotel room after the affair was over. Then, when he broke the seal, he found nothing in the envelope but a few newspaper clippings.

"I had a whale of a time anyhow," McCurdy said.

That seems to have been very true, and was perhaps the reason why he and his brother Lucian missed the boat which was to take them back from Havana to Key West. He confesses that they were guests at a party given by a son-in-law of Pierpont Morgan, and that they were having a cosy little celebration at a fashionable seaside club, when he was astounded to see his boat pulling away from the wharf. Since McCurdy had a very important engagement to keep in Tampa, he knew that he must get onto that boat. McCurdy's newfound friends were very sympathetic when they learned of his predicament, and promptly grabbed up his packed bags and loaded them into their launch which was tied up nearby. With commendable speed the launch struck out across the water in pursuit of the big steamer bound for Florida, and eventually came alongside. All their shouting and frantic gesticulating to the captain were of no avail, however. "We can take no one on board now!" the captain said with some indignation.

McCurdy noticed, however, that at the rear of the boat and not too far above the waterline, the two portholes which received the vessel's load of coal had been left open. McCurdy got his friends to bring the launch alongside, and taking a last fond look at the shining white of their newly purchased tropical suits, he and Lucian caught the lower lips of the portholes and clambered in while the captain shrieked curses at them from above. Inside, the hold was so dark that McCurdy could see nothing at all for a moment. In fact, for a while he was not sure that Lucian had been able to get in. But Lucian was there sure enough and as their eyes gradually became accustomed to the gloom, they saw a ship's officer picking his way gingerly over the coal and the gratings toward them. The two stowaways were seized and paraded up on to the deck where the captain was gathering a full head of steam. The effectiveness of his anger was impaired somewhat by the tittering of the rapidly assembling passengers. What the coal had done to the two white tropical suits was quite delightful for them to see.

McCurdy did his best to explain, but the captain did not give him much chance. "You thought that I refused to stop to pick you up simply because of the inconvenience it would cause me," he said. "You are wrong. I will have you know that you are illegally aboard since you have not passed your quarantine examination. No passenger is allowed aboard this ship without a proper doctor's certificate and you ought to be well aware of that fact. This is a serious breach of the law and you might as well know that you will answer for it! You will have to be placed in confinement until the authorities in the United States decide what to do with you!"

But once again McCurdy's fame was to stand him in good stead. The ship's doctor, who had followed the reports of the McCurdy flight with glowing enthusiasm, now stepped forward and said quite calmly: "You are mistaken, Captain. I examined both these men this morning in Havana. They are both quite all right."

It was a good lie, nobly spoken, and the captain had no choice but to accept it as the gospel truth. McCurdy arrived back in Tampa in time for his appointment and to the tumultuous cheering of thousands.

12

Washington Watches

But the acclaim which was McCurdy's that year was not entirely due to his record-making flight to Havana. He had been already regarded as one of the finest aviators of America, if not of the world. The accident rate for other barnstorming pilots in those days was frightfully high, and when McCurdy continued to fly month after month, and under all kinds of field and weather conditions, without mishap of any kind, it became increasingly clear that he had something more than luck in his favour. His insurance against accident was plainly the result of skill, cool thinking and eternal vigilance.

Yet for all his reputation for caution, McCurdy could always be depended upon to put on a rousing good airshow. Other pilots of his day had a tendency to rely upon daredevil feats of aerial gymnastics, but McCurdy's acts often gave the spectators something to think about. And many of his unique exhibitions were quite evidently designed to let the audience peer into the future.

One of these unique acts was called "Battleship Bombing," and for this part of the show McCurdy used stage scenery that, like the settings which Shakespeare once used, demanded a great deal of imagination but very little money. His "battleships" were mere patches of cloth, and his "bombs" were oranges.

The *Mobile Register* said in one of its Sunday editions that winter:

"McCurdy blew up a battleship Saturday. That is, he would have blown one up if the oranges he threw from his aeroplane had been filled with nitroglycerine, and if the 12-foot squares of canvas lying on the ground in front of the grandstand at the aviation field had been the funnels of a man-of-war.

"When the time to begin the afternoon's series of exhibitions arrived, the aviators, Post and McCurdy, were themselves afraid that the wind would prove too puffy to permit of spectacular flights. In order to ascertain the state of the air, two trials were made—the first a straightaway flight by Post, and the other a flight by McCurdy, who ascended into the higher air currents of the upper levels.

"Post's trial flight, made in the 25-horsepower biplane, carried him straightaway down the field and ended near the southern end of the enclosure. McCurdy's first flight, made in the 50-horsepower Hudson-Fulton Flyer, took him up to a height of about 600 feet. At this elevation he sailed around two great circles as prettily as could be desired. His landing was, like all the others made by him during the week, a graceful performance that won applause.

"Then came the principal feat of the day—that of throwing from an aeroplane imaginary bombs at the imaginary funnels of an imaginary warship. This was performed by McCurdy.

"Ascending in the Flyer, the young Canadian who for a week had been thrilling the thousands of persons who attended the aviation meet, flew down the field and returned, travelling at a height of about 100 feet, and at a rate of at least sixty miles an hour.

"In the field a short distance in front of the grandstand, there were two twelve-foot squares of canvas, about 75 feet apart, and these represented the funnels of a battleship. Suspended from McCurdy's neck and resting in his lap was a narrow basket in which were half a dozen oranges.

"As he passed over the imaginary man-of-war, he dropped one of the oranges. The small yellow globe was carried forward by the momentum of the biplane more rapidly than the law of gravity drew it toward the ground. How, under the circumstances, McCurdy could tell when to drop the orange in order to cause it to strike the canvas, no one on the ground could form a clear idea; but the fact remained that the orange struck the edge of the forward piece, and with a force that tore it to pieces.

"On the second passage over this supposititious battleship, McCurdy misjudged his speed a trifle, and the little spheroid struck the grass about twelve feet from the forward canvas. The orange landed on the deck of the imaginary vessel.

"On the third passage, the aviator's aim was more accurate, the orange landing in the center of the square chosen as the mark.

"The fourth and last attempt resulted in a miss of about four feet as far as the imaginary funnel was concerned, but the make-believe missile of destruction landed squarely in the imaginary upper works of the imaginary war vessel, and the explosion that followed would have been terrific if it could have been heard."

But even had McCurdy been able to drop oranges with the accuracy of a Daniel Boone, the achievement would have been a much smaller one than the marvel which he performed a few weeks later at Palm Beach.

The date was March 6, 1911, a scant month after his triumphant return from Havana. Palm Beach was just seeing the end of a marvellously successful "Aviation Week," and as so often happened at such events, the high point of the exhibition was provided by McCurdy. And as so often before, it was the freshness of the idea rather than any act of daring which made the crowd take notice. But this time, the whole nation was to take notice.

Some time during that week, McCurdy co-operated with a DeForrest wireless engineer by the name of Percy G.B. Morriss to set up a temporary wireless transmitter and receiver at Bethesda Park, which was the field where the meet was being held. After the installation had been tested and adequately proven, the two men next devised a light-weight wireless apparatus and attached it to the Curtiss biplane which McCurdy was flying that week. Again the device was tested, but it was discovered that while the wireless on the plane could transmit perfectly, it was almost impossible for the operator to make out any messages being received so long as the motor was going. The roar was too much.

The cure was a trifle ridiculous, but quite effective. It had already been agreed that Morriss would fly with McCurdy and operate the wireless, so McCurdy simply covered his partner's earphones with two wet towels and bound the strange cap tight with a big ball of tire tape. When he was finished, Morriss looked as though his entire head was in a plaster cast.

The crude arrangement insulated the engineer's ears against the deafening racket of the engine, and on the final afternoon of the meet, with the plane sailing out over the Caribbean at an altitude of 2,000 feet, Morriss made a clear contact with the Key West transmitter as well as with the temporary one they had just left behind them at Palm Beach. An even greater thrill came when Morriss picked up a steamer of the United Fruit Company as it left Panama bound for New Orleans.

It was the first time that the wonder of wireless had been combined with the wonder of flight, and if many of the fun-loving holiday folk of Palm Beach failed to see the dramatic import of the combination and forgot it a day or two later, there were some of the quietly great minds of the country who did not. Twelve days after the wireless flight with Morriss, McCurdy was in the nation's capital demonstrating his Curtiss biplane before a sparkling assembly of War and Navy officials, and numerous representatives of foreign countries.

In an article of March 18, 1911, the *Washington Post* described the event as follows:

"With representatives of the War and Navy departments and military and naval attachés from a number of foreign legations here looking on, the aeroplane just purchased by the War department to be used in the Mexican border manoeuvres was given its initial and successful demonstration here early yesterday in Potomac Park. The machine, which is what Curtiss designates as his 'military' type, was given a thorough tryout by J.A.D. McCurdy, who made two ascents, each time circling the big lot to the south of the State, War and Navy building. So close did he sail to the building that the whir of the war-bird's propeller was heard plainly in the office of the Secretary of State. Also he circled the Washington monument.

"In all, McCurdy was in the air about eighteen minutes. His first flight covered about 250 feet. McCurdy then glided to the earth where the machine was inspected by the aviator and the small crowd. He then started the motor and began a second cloud-soaring expedition. This time he went far above the Washington monument.

"The aeroplane was taken from the exposition building, North Capitol and M streets, where it has been on exhibition at the aeronautical show all week. Following the test the aeroplane was returned to the show."

The newspaper then goes on to name the foreign officials who were present, and the list is one which must make interesting reading to students of American foreign policy. The *Post* reported that of all the spectators present, the Japanese were the most intrigued.

"Japan has already purchased eight of the new machines and is negotiating for several others," the article concludes.

But to McCurdy, the spectator whom he was most delighted to see wasn't a military man at all, but an old friend he had met in Washington years ago. This was David Fairchild, who was now married to Daisy, the younger of Alexander Graham Bell's two daughters—the same Daisy who had once been an honoured guest aboard the Huck Finn houseboat.

There was a lovely reunion with the Bells that night in Washington, and when it was all over, McCurdy decided that he was just a little homesick. The great republic had helped him soar to his new fame and had treated him like a young god. But Canada was still home to him, and he knew that he would soon go back.

13

The "John the Baptist" of the Royal Canadian Air Force

The summer of 1911 saw McCurdy back in the land of his birth and more determined than ever to make Canadians an air-conscious people. In August, he staged an air race between Hamilton and Toronto, and for his partner in this venture, he had a young southerner by the name of Charles F. Willard who had flown in southern California between Los Angeles, Pasadena and Alhambra. While several flights had been made between the two Canadian cities, none of them had been as yet non-stop, so McCurdy thought that the event which he had planned should be of outstanding interest to the whole country.

Briefly the plan for the race was this: Willard and he were both to leave the aviation field four miles east of Hamilton, but they would take different courses to Toronto. McCurdy would fly directly over Lake Ontario, and Willard would follow the shoreline. They were to meet at the waterfront in Toronto.

Willard left first. Thirteen minutes later McCurdy took off. Willard met adverse weather conditions which tossed him about rather badly, while McCurdy, who was generally thought to have selected the more hazardous course, had a very peaceful flight over the lake. Both men registered 35 miles for the flight. McCurdy made his in 36 minutes. Willard took 43 minutes. It was the fastest time that had ever been made between Toronto and Hamilton, and should have given Canadians a graphic indication of the advances which aviation had made since that first day just two and a half years previous when McCurdy had lifted his *Silver Dart* from the ice at Baddeck Bay.

But Canada seemed only mildly interested, and the substantial citizens were apt to be frankly critical. It was not until May 15th, of 1914, almost three years later, that the first passenger was flown between these two cities. That passenger, incidentally, was Lou E. Marsh, a sports writer for the *Toronto Star* who also had a coast-to-coast reputation as hockey and boxing referee. He flew with Theodore McCauley in a hydroplane named *The Sunfish* which was owned by W.A. Dean of Toronto. The trip was made in 32 minutes and 17 seconds, in a direct line over the lake. But even at this late date, Canadians thought such an undertaking foolhardy and hazardous, so a motorboat was chartered to follow the course in case of emergency. But the plane sped along at 72 miles per hour, much too great a speed for the boat, which was soon lost to sight.

The caution and complacency which seemed almost a Canadian instinct in those days was soon to vanish, however, for just three months after Lou Marsh's brave hop, the Empire was at war. Overnight the grim eventuality, which McCurdy had so often visualized in his exhibitions and in his speeches about the future of flying, had come to the world. Fired with a young man's patriotism, he lost no time in getting to Ottawa where he succeeded in having an interview with Sam Hughes, then the Minister of Militia and Defence. Hughes sat impatiently drumming his fingers on his desk while McCurdy did his best to tell him what he thought the aeroplane could do for the defence of the country in this hour of her greatest emergency. But the interview was brief, Hughes snapping it short with a speech which rang in McCurdy's ears for many years after.

"The aeroplane is an invention of the devil and will never play any part in such a serious business as the defence of a nation, my boy!"

McCurdy flushed, and said quietly: "I am sure that you will live to regret those words, General Hughes."

But McCurdy was not to be shaken and never passed up a chance in those early war months to be an evangelist for the cause of military flying. He was an excellent speaker, and audiences listened to him eagerly wherever he went. A short time after he had his sad interview with Sam Hughes in Ottawa, he said in Toronto:

"There is no reason why we should not have a squadron of at least twelve planes here in Canada to assist the militia. The day is gone when the artillery shoots over a hill at the enemy. The modern method will be to get down behind a hill and let the aeroplanes tell the tale of what is hit or where the shots go. What we want the government to do, of course, is to send some machines over to the Allies and, in the meantime, to maintain a corps of our own.

"An aeroplane would be absolutely safe from the guns of the enemy at a distance of 5,000 feet because it takes so much more power to send a shell in a vertical direction, and at that distance a thirty-foot spread aeroplane would be nearly invisible."

In another part of that same speech, there are some interesting observations about the types and performances of the planes which were available that first winter of the war. "The main difference between a naval and a military machine is that the military is a faster climber," he said. "It is not impeded by the weight. The military machines will climb 4,450 feet in ten minutes with a pilot, observer, gas for five hours and 144 pounds extra weight. A naval machine would only climb 3,000 feet in the same time. The faster climber would be easier to manoeuvre and could get up on top of the slower machine from which position it would be easy to put it out of business."

Referring to one of the earliest raids of the war, that of 34 British aircraft on Belgian towns in German possession, McCurdy had this to say: "A bomb raid has a two-fold object. One is to get information, which, by the way, is the chief function; and the second is the moral effect where bombs are dropped, say on towns. From an economic standpoint there is not much loss, but there is no doubt that bombs dropped on a camp in the dark or the early morning would have a strong disorganizing influence and I believe that the British raid will have its effect. If we can disturb the morale of the enemy, that would in itself be a great victory."

It was a prophetic speech indeed, and an amazingly wise one to come from one so young. McCurdy was only twenty-eight that year. Perhaps it was as much his own youth as that of the aeroplane which made his fellow Canadians hearers of the word only. As time went on, it became increasingly evident that whatever McCurdy would do to further the idea of an air arm for the Canadian military effort, he would do without government help.

War makes strange bedfellows, but it makes even stranger politics. And it was a curious clause of American war policy which was indirectly responsible for giving McCurdy a chance to start Canada's first aviation programme. When Britain declared war in August of 1914, the whole Empire was at war, Canada included. But the United States was not a belligerent—it was indeed belligerently determined that it would remain at peace. There was no national reluctance, however, to make a few of war's necessities for the fighting nations if the price was right. But these necessities were rather clearly defined and weapons were not included.

What was the aeroplane?

Conscientious Americans considered the problem from every angle and finally decided that a plane without ailerons could certainly not be considered an instrument of war. And with that question in national ethics satisfactorily resolved, several American factories went to work immediately to assemble planes for shipment to Britain. Glenn Curtiss made them by the hundreds in a new plant at Buffalo. His old factory at Hammondsport turned out the motors to power them.

So it was that from the great republic where hate of war was a national religion, shipload after shipload of these planes and motors went across the water. Britons received them gladly. They had no time to ponder the mystery of the American morals which sent the planes over without ailerons. Besides, they knew exactly where the ailerons were to come from. They would come from one of the inventors, a burning young man by the name of J.A.D. McCurdy, who now had a plant of his own in Toronto big enough to turn out the missing parts as quickly as they were needed.

That plant was situated on Stone Street on the site later occupied by the John Inglis Company. The factory was called Curtiss Aeroplane and Motors Limited. But while Glenn Curtiss was still a very close friend of McCurdy, and while the ailerons were being built to fit the Curtiss planes, the American airman had no financial stake in the venture. McCurdy was president and proprietor.

Within the space of a few short days, the aileron factory had very shortly grown into a flying school as well, and a study of the manner in which the McCurdy Flying School was established provides disturbing proof that the Americans were not the only people capable of peculiar politics in wartime. Officially, the Canadian government was as determined as ever that there should be no national air arm. Aviation may have had its friends in the Cabinet, but when Sam Hughes roared his tremendous "No!" to the idea, no one dared to argue with him. As Minister of Militia and Defence, the old Boer War veteran now had a power and a prestige that rivalled even that of the Prime Minister.

It was the British who really made the McCurdy Flying School possible. With the Canadian militia department giving the plan little more than permission, an arrangement was made for the recruiting and training in Canada of candidates for the Royal Naval Air Service. McCurdy's school was to remain strictly a private school, and the government was repeatedly demanding press room to make sure that the public knew this. Only the preliminary training was to be completed in this school, and when the enlisted man had earned his aero certificate, he was to go to Britain to finish his training. The enlisted man paid his own fee of $400 for his instruction, but if he completed his term satisfactorily and was granted his certificate, there was a provision under which the British government refunded $375 of that fee as well as giving him an unstated allowance for uniform and kit and a second class passage prepaid to England.

Shortly after the Royal Naval Air Service had concluded its agreement with McCurdy, the Royal Flying Corps also decided to recruit Canadians under the same plan.

On the surface this was a fine turn of business indeed, but McCurdy discovered to his grief that it was apparently a tradition for the two British services to cold-shoulder each other. At first McCurdy could not believe that there was anything serious or lasting about this unusual rivalry, but when he found that officers of the Royal Naval Air Service would not even speak to the more recently arrived personnel of the Royal Flying Corps, he concluded that the only thing to do was to incorporate two separate flying schools.

The school to accommodate the Navy men was established on Toronto Bay and used Curtiss Flying Boats. For the Royal Flying Corps, Curtiss biplanes were used, and the Canadian government was finally persuaded to give permission for the construction of a flying field at the Long Branch Rifle Range. The government went even one step further in its cautious sanction of the idea and decided that an Order-in-Council which had been passed in September of 1914, and which made it an offence to fly an aeroplane within twenty miles of any Canadian city, would be set aside so that McCurdy's trainees could train in the vicinity.

Horses and scrapers were hurriedly assembled to level off the butts of the old range, and on April 30th of 1915, both schools were in full operation.

Far from being a daredevil's prank, it was suddenly a noble thing for one to frolic in the heavens now, and Toronto papers of that year gave dozens of detailed accounts of the glamorous happenings at the flying school. The following account is from *The Globe*, May 15, 1915:

"With their eyes turned toward the east, as they will be on a later summer morning when they speed their flight from England to the battle against the Huns, Canada's first airmen-to-be sail over Toronto Bay every morning before the city is well awake.

"They are early birds. The roll is called at 5 a.m. A calm atmosphere is desirable for teaching the novice, and the winds are still asleep for some hours after sunrise. At the same time one has but to meet these young airmen to come into touch with an enthusiasm and earnestness for their calling that would bring them from their beds at a still earlier hour, were that necessary. A representative of *The Globe* was vividly impressed by these qualities when he accepted an invitation to accompany the airmen to their Island base.

"The rendezvous was the foot of Spadina Avenue, where a gasoline launch is moored. At a quarter to five in the morning the city lamps had just been extinguished, when a light shone out in the grey dawn from a small houseboat on the shore, a door opened, and a man and a dog appeared. It was the first move in the day's game. The boatman who accompanies the flying men in his gasoline launch had been awakened by telephone by Instructor Pierce, and with his good dog Nick was getting everything ready. Then, just a few moments before 5, Capt. McLaughlin, Ottawa, strode leisurely up, having already breakfasted. Then came student No. 2 on a bicycle, with Instructor Pierce, slim and neat of build, a moment after. It was now 5, and the words 'All aboard' were about to be shouted when a taxi drove up. A young man with a bottle of milk in his hand emerged and dashed down to the wharf. It was Lieut. Hay of Owen Sound, who had had to come from a longer distance than his fellows or had overslept and had picked up a taxi and his breakfast on the way down.

"It was with a sigh of relief he found he was in time. But he was not last. Another taxi drove up and out leaped the tall Peberdy in slouch hat, Lieut. Homer Smith in military uniform, and a few others. Courtrage, of Brantford, was unable to come and Cyril Day was the only absentee. Perhaps he had come down earliest of all and read a notice pinned on the houseboat door, 'Back at 10.30 p.m.'—a very interesting sign indeed to meet the eye of an ambitious and early-rising aviator, as Mr. Peberdy remarked. The notice, however, was one that had been posted the previous evening.

"It was now 5.10. 'All aboard!' was the call and off we went to the Island sandbar. In the grey, calm morning a barque was making sail for the harbor and sea birds flew overhead. All eyes studied the sky.

'Won't we give that schooner a fright when we fly over her,' spoke the irrepressible Lieut. Smith.

"The boat touched the shore and we leaped out on the sand and entered the hangar. The next stage in the flying was reached—the awakening of George, the mechanic, who slept keeping watch near the flying boat. A few handfuls of sand with sundry odd ends of rope thrown over the wooden partition brought quick shouts from the interior and told that the aim had been true, and in a moment George, thus unceremoniously awakened, emerged and set the engine agoing. This was an operation that required patience till the engine warmed up. There was pity for George, who had to exert himself thus strenuously before breakfast. He got lots of advice, however, to 'put some pep into it.' They are all humorists, these students.

"'All aboard' again was the call from Instructor Pierce, now clad in a blue coat, with a balaclava helmet on his head and rubber shoes on his feet. Lieut. Smith took his seat on the boat, which slipped out over the lake, slowly at first, then faster and faster as the chug, chug of the engine increased in intensity. Sixty miles an hour is the speed the boat is capable of on water and about 65 miles in the air. All gaze after it. 'Thirty,' cries Peberdy, as the boat travels faster. "Forty.' 'Now he's in his swing.' 'Look at her tail rising.' 'She's on the surface of the water.' 'There, she's in the air now.' 'She's going up by bounds, my, that's the highest flight yet.' 'He must be 300 feet up.' 'Compare them with the C.P.R. building.' Then the flying boat swerved around for home, descending gently to the water. Instructor Pierce brought her level a few feet above its surface and she dropped down with a slight splash. A moment more and Lieut. Smith stepped ashore. He had been away only six minutes, but 'that was the greatest sport,' he said. 'How did you like it going up?' he was asked. The Lieutenant admitted it scared him a bit. It felt like someone pouring cold water down his spine.

"G. Gooderham was the next to fly. This flight was even higher than the preceding one and of wider sweep, and at one time the boat was invisible in a cloud. Then down it swept to the water most gracefully with scarcely a splash. Mr. Gaele was the next. He soared and sank again to the water beyond the breakwater. Was he to soar again to cross that barrier? It would be a close shave, but it looked like being attempted. 'No fear of that,' said Peberdy, or was it George? 'Pierce is a Conservative and never takes chances. He'll loop the loop if need be, but safety first is his motto—a good one, especially in an instructor.' And the speaker was right.

"So the morning's round went on, each student in turn having about ten minutes' training. They will have had 400 minutes each in all when their course is complete, and this will cost them a dollar a minute. The finishing touches will be given in a biplane at Long Branch. Then the students will be ready to pass the tests that will admit them to the British Naval Service."

The test which a student had to pass, before he was given his aero certificate, will be of special interest to pilots of today. The candidate had to make two distance flights, consisting of at least 16,404 feet (five kilometres) each, in a closed circuit, without touching the land or water. These flights were made in "figure eights," and at least eight "figure eights" had to be made. Then the aviator had to make an altitude flight of at least 328 feet, and to descend from that height with his motor cut off. His landings had to be clean, and in the distance flights he had to come to rest within 164 feet of a designated point. On these tests he had to be alone in the machine. On the average, three to six weeks of instruction were required to make a man competent to try his examination. As to the actual time he was required to log in the air, no rigid regulations were laid down. When the student paid his $400 fee, however, he was entitled to 400 minutes of actual flying. But if he happened to be the kind of man who learned flying easily, he might very well be permitted to try his examination long before that. This was particularly so as the summer wore on, and the war across the water began to go badly for the Allies.

The following is a piece of news which appeared in the *Toronto Star*, September 8th of that year:

"AN AVIATOR IN TWO HOURS

"'You'll do,' said pilot to pupil after just three flights. The phenomenal record of learning to manipulate a flying boat in 118 minutes' experience goes to Edward Blumfield Waller, of 192 Jarvis Street, a student of the Curtiss Aviator School. Young Waller began to fly only last Wednesday, and passed his final test in the hydroplanes yesterday. On his third flight, which was made early last Thursday morning, Waller noticed the pilot fixing his goggles with both hands, and was surprised to learn on reaching the hangar that he himself had been in complete control of the machine during the entire flight.

"'You'll do,' was the comment of the pilot on landing. He made two flights Wednesday, two Thursday, two Saturday, three Sunday, and passed his test on Monday, which consists in shutting off the engine and darting to the water from a height of 300 feet, besides flying a given course.

"He was an employee of the British-American Insurance Company for two years prior to taking the course in aviation. He formerly attended Parkdale Collegiate."

But in spite of the speed with which it was required to produce, and in spite of the temperamental motors and the crudity of the craft which they had to fly, McCurdy's pupils had a remarkable record for safety. In all the months that the school was in operation, not a single student was killed or injured while training. Nor was there any complaint from the British about the quality of training which the McCurdy graduates brought with them to the finishing schools over there.

Overseas, the grim clutch of death was not so easily denied, and in early November of that year, just as his parents were getting a birthday package ready to send to him, Canada received news of the first tragedy to come to its gallant airmen. The man who was killed was Lt. Douglas Hay of Owen Sound, the same lad who had amused the *Globe* reporter the previous May by coming down Spadina at top speed at five in the morning, drinking his bottle of milk on the way.

For all the school's amazing efficiency, and its undoubted contribution to the Allied cause in Europe, Sam Hughes and the Canadian Cabinet were still adamant in their refusal to consider a separate Canadian air force. Criticism of flying was not confined to orthodox politicians, however. The good city of Toronto has always had more than its share of the type of clergy who would like to legislate men into the Kingdom of Heaven, and in August of 1915, the Reverend Dr. Moore, a Lord's Day Alliance official, got his name into the papers by declaring it a sin to watch aeroplanes on Sunday. "Sunday aviation tests should not be made before public spectators," he said. He was very cautious, however, about venturing an opinion as to whether or not it was proper to engage in flying of a military nature on the Sabbath.

But such comments were getting rather faint and far between now, and criticism of an entirely different sort was gaining weight rapidly. Just six months after the first student had taken to the air over Lake Ontario, public condemnation of Canada's official frigidity toward flying had so grown in its severity that the issue was being discussed in editorials right across the country. Said the *Toronto Star* on October 9th:

"When a man enlists to serve his country in a battalion or battery, he is immediately put on the country's pay list and is paid for his time while he is in training. When a man, however, undertakes to prepare himself as an aviator, he does so at his own expense. He not only is obliged to give up whatever position he previously held, but his necessary expenses while in training are, if he is boarding, not less than $15.00 a week, for which he gets absolutely no reimbursement from the Government.

"It appears that one cause of this discrimination against aviation students is the fact that the Canadian Department of Militia has hitherto failed to recognize aviation as a branch of the Canadian army, although it has been demonstrated since the first day of the present conflict that for modern armies an aviation branch is absolutely essential. The aviators are now described as the eyes of the army.

"Ottawa Doesn't 'Know' Them

"Canadian aviation students who get their diplomas are given an opportunity to enlist in British flying services, either in the army or the navy, as they may select, but they get no recognition from the Canadian Government. Yesterday when Mr. J.A.D. McCurdy, at Ottawa, asked the Prime Minister for assistance toward the expense of transporting student airmen to Bermuda for the winter, the Premier is reported to have replied that such a grant would be without precedent, as the airmen were not being trained for a Canadian service. This is typical of the attitude of the Government toward this branch.

"The situation furnishes another example of where individual citizens are leading the Government in service to the Empire, instead of the Government itself giving the lead."

In the same paper, the following letter from a student's irate father shows the injustice of the situation even more pointedly than the editorial:

"To the Editor of the *Star*: It is fair to assume that the public generally, as well as the Militia Department at Ottawa, do not appreciate the injustice being done to those who are preparing themselves for aviation work, or there would be a change in conditions. A few facts will make the situation clear.

"In the first place, a rigorous examination is necessary, and only those of excellent physique and with splendid eyesight can qualify, and this plus the best of references. Provided the would-be aviator passes said examination, he or his parents are required to pay $400 for tuition. Then the student spends his entire time from early morning until night in order to take his turn, first with the hydroplane at the Island and then later at Long Branch with the biplane.

"Below will be found an accurate computation of the expense which the course means to a boy whose home is not in the city, allowing nothing whatever for the loss of the position he has previously held, with remuneration it carried. It must not be forgotten too, that rainy or windy weather delays the work of the student aviator, but his expenses go on just the same. The positive and unavoidable cost is as follows:

Sleeping room at the Y.M.C.A per week	$4.00
Board at restaurant 35 cents each meal, week	7.35
Streetcar fare to boat, once each way daily, per week	.70
Boat fare to Island and return, twice daily, per week	1.40

Boat fare across the Lagoon, twice daily,
per week 1.40

Making a total of $14.85

to say nothing of laundry and other incidentals which must be arranged for. These figures are actual and cannot be disputed. Now, an ordinary soldier, as soon as he passes his medical examination and begins his drill, draws, I understand, $1.10 per day, with bed and meals included. The aviator provides himself with everything, assumes a very much greater risk, and gets absolutely nothing until such time as he obtains a certificate from Ottawa. Further than this, weather conditions will soon necessitate the Government sending the boys (soldiers) to a warmer climate, probably Bermuda or Victoria, British Columbia, they (the Government) paying transportation expenses.

"It seems to me that, to say the least, the treatment accorded the soldier-boy-aviator is unjust and unworthy of Canada. It is fair to state that on arrival at England and when the aviator satisfies the British officials, he has $375.00 of the $400.00 paid in refunded, together with an allowance for uniform of $250.00. I feel sure that parents who give their sons in the service of the Empire are not asking too much when they suggest that these facts should be fully weighed. The aviators and their parents ask only a fair, square deal, and, in their opinion, it is up to the Government to correct these conditions. "

Then, just as the cloud of criticism was gathering toward a storm point, the British Army announced that it would no longer use Canadian flying schools to train personnel for the Royal Flying Corps. Canadians who wished to fly in the war would take all of their training in Britain. Within a few weeks the British Navy followed suit by announcing that henceforth all of its flyers would take their entire training in Britain.

Since Ottawa still stubbornly refused to consider the creation of a Canadian air arm, there was no further use for McCurdy's Flying School, and in November he announced its death. He had graduated 300 of the finest airmen in the Empire, and among them were names which were to become nationally known in the years to come. Among these were such famous men as Raymond

Collishaw, W.A. Curtis, and Robert Leckie, all of whom later became Air Marshals. And down in the States, another flyer who had learned his trade with the associates at Hammondsport was also beginning to make a name for himself. His name was Henry H. (Hap) Arnold who was later to become General of the United States Army Air Force.

But there were 300 other students whose courses were not completed, and McCurdy refunded their tuition fees and bade them good-bye. Rumours to the contrary, he had made no fortune with the school. It had, in fact, cost him dearly. But his regrets were for his failure to convince an immovable government that the air age had come to Canada as it had to the rest of the world.

Next spring he once again made a trip to Ottawa and this time was successful in speaking to the Prime Minister, R.L. Borden. McCurdy was still doggedly clinging to the idea of an officially backed air training programme for Canada. He was not sure any longer just how this might best be brought about. But his graduates of the past year had acquitted themselves so gloriously flying with the British Forces that he thought perhaps the British might be led to reopen the Toronto flying schools. Or failing that they might be able to influence Ottawa to open them as Canadian schools.

The Premier gave him the following letter of introduction to Sir George Perley, the Canadian High Commissioner in London. McCurdy treasured the original of this letter to remind him of the days when he was fighting a lonely and losing battle for a military flying body in Canada:

Prime Minister's Office, Canada Ottawa, April 11th, 1916

Dear Sir George Perley,

This will serve to introduce to you Mr. J.A.D. McCurdy presently of Toronto who is interested in the manufacture of aeroplanes as well as the training of Canadians for service in the British Air Service.

Mr. McCurdy informs me that his Company has completed a number of aeroplanes for the British Government and a considerably larger number are now under construction.

During the past year the school at Toronto has trained about three hundred Canadians for the British Air Service all of whom as I understand have acquitted themselves very well in the further training which they receive in the United Kingdom.

He has suggested to us the establishment of a Canadian Air Service and we have that subject under consideration at the present time.

I would be glad if you would give Mr. McCurdy any necessary assistance in obtaining an interview with Lord Derby whom he wishes to see and also any other official whom he might desire to see.

Mr. McCurdy has lived for some years in the United States but he is a Canadian, being a member of a wellknown Nova Scotia family.

Believe me,
Yours faithfully,
R.L. Borden

As a result of Mr. McCurdy's visit to England the Imperial authorities took over the School and in turn handed it to the Royal Flying Corps. A staff from England proceeded with the training and opened up many additional schools, in which training was continued until the end of the war.

McCurdy got a highly secret order for the manufacture of a special military bombing machine for the British Navy. The specifications called for twin motors on this plane, and a top speed of 80 miles per hour. McCurdy's factory built the plane, thus turning out the world's first twin-engined bomber. The first model exceeded the required speed by a full six miles per hour, and the British were so delighted with it that they immediately ordered more of them. By the time the war ended McCurdy had turned out thirteen of the twin-engined craft besides a considerable number of JN-4's, a craft which later became the official training plane practically everywhere in the Empire.

It was not until 1922 that a new Canadian government, by Order-in-Council, finally established the aviation arm which McCurdy had first proposed in 1909. The new force was first called the Canadian Air Force, and in 1924, when the idea was ratified by statute, the name was changed to the Royal Canadian Air Force.

The event was something more than an important bit of history. It was a victory for youth and a new idea over the solid men who were unalterably convinced that to be safe, proper and good, one must be changeless. And no historian would ever know the long pain of that struggle as McCurdy did.

14

The After Glories

It has been the intention of this book to tell the story of the young McCurdy, rather than that of the man after his days of pioneering were safely behind him and his wild and lonely dreams had finally come to earth. Indeed, it might seem to some that the list of accomplishments of this slight Cape Bretoner's first thirty years was so remarkable that little lustre could be added to it. He had not only been the first Britisher to make a controlled flight, but he was the ninth man in the world to fly a heavier-than-air machine. He had contributed largely to the invention of the aileron, the tricycle landing gear, wing dope and pontoons. He had been the first man to fly a plane powered by a water-cooled motor; had made the first great ocean flight; had made the first flight with skis; had been the first to demonstrate the possibility of bombing from the air. He had given flight control a dramatic new meaning by flying the first figure eight; he had been the first pilot to send and receive a wireless message while aloft.

And he, more than any other, had been responsible for the idea of the Royal Canadian Air Force.

But the end was not yet, and the fact that this biography dwells so briefly upon the achievements that came after by no means indicates that these were of such lesser importance that they do not merit the space. To honour them justly would require enough chapters to make another book, but that would have to be a book of an entirely different nature.

Soon after the ending of the First World War, McCurdy married. And while his manager-brother Lucian did not actually manage the romance, he was certainly instrumental in getting it started. It was in Toronto on a Christmas Eve, and McCurdy had been holidaying with his brother, who was already married. Here he met a friend of Lucian's who could talk of nothing else that night but of the beautiful woman whom he claimed as his sweetheart.

"You ought to see her!" he said. "She's got everything! I want you to meet her!"

Undoubtedly, the young man thought that his sweetheart would be tremendously impressed if she could only see him on such intimate terms with a national hero. Actually, when the time came, she wasn't at all impressed. Unthinkable as it was, she wasn't even sure that she knew what this so-called hero was famous for. It was McCurdy who was impressed. He discovered that, for once, Lucian's friend had not exaggerated.

Her name was Margaret Ball of Woodstock, Ontario, and then a pupil at the ultra-fashionable Margaret Eaton School of Literature and Expression, and a daughter of Oxford County's Crown Attorney, R.N. Ball, K.C.

"She was perfectly beautiful," McCurdy declares. "Lucian's friend was right. She had everything. And it certainly seemed a great pity to let him have her without a contest."

When he tried to date her, however, Margaret was distressingly cool to the idea. Even when her friends tried to tell her how wonderful a man he was, she was still but mildly interested. She did finally consent to go out with him afterward, but only with a chaperone. And true love did eventually come along.

The McCurdys had two children. Dianna, the older, became the wife of Captain Philip Haddon, of the Royal Canadian Navy, and they had three children. A son, J.R.D. McCurdy, became a prominent business man in Montreal.

After the Armistice, McCurdy naturally turned to thoughts of manufacturing aircraft for civilian use, but the transition was not an easy one. Interest in aviation had soared during the last years of the war, but the very idea of war meant daring and the constant threat of death. Now with the coming of peace, the world wanted desperately to have done with danger and the aeroplane was still considered to be a very dangerous machine.

It wasn't until the mid-twenties that spectacular trans-ocean flights convinced the world that it was surely only a matter of time until the aeroplane would become a great new travel medium.

McCurdy had always believed this, and now with the public slowly but surely overcoming its fears and prejudices, he made plans of his own to take an important part in the new air age. In 1928, he formed the Reid Aircraft Company and established a plant at Montreal for the manufacture of various types of planes. The year following, the company was merged with the Curtiss Company and chose for its new title Curtiss-Reid Aircraft Ltd. It was a name which was soon to be known the world over, and McCurdy was its president.

But even as a mighty executive, McCurdy was still the adventurer, and during the early 'thirties he made two trips abroad which had more of mystery and international intrigue in them than a piece of pocketbook fiction. In September of 1931, Japan struck at Manchuria and, early in the next year, Nipponese bombs began to fall on Shanghai. Feeling in the Western world was strongly on the Chinese side and there was even agitation in some quarters to go to war over the issue. McCurdy had his own idea, however. He made the suggestion to some influential Chinese in the United States one day that there were many fine military flyers in Canada who would doubtless be glad to go to China to fly against the Japanese if they could be assured of suitable support and co-operation. He suggested further that he might be willing to help organize those flyers.

Somehow the suggestion got through to China itself and a few weeks later, McCurdy was there, too. Here, with the utmost deference and the most mysterious of manoeuverings, he was met by a very lovely Chinese woman who told him that she would take him to a meeting of the Chinese military men who would be glad to listen to his plan for helping their stricken nation. But the meeting place of these leaders was a closely guarded secret, and if McCurdy wanted to see them, he would have to allow himself to be blindfolded.

McCurdy submitted, and finally found himself in the smoky presence of some thirty of China's war lords. They listened carefully and politely to his plan, and finally sent him away telling him that they would consider the matter in his absence and would let him know their decision.

The plan came to naught, probably because of the jealousy which existed then between the lords themselves, but McCurdy had an introduction to oriental diplomacy which he would never forget.

Forces and new ideas were on the march in Europe, too, and 1933 saw McCurdy in Britain and Germany as the honoured guest of several of the new aircraft manufacturers. In Germany, McCurdy was given an escort and a special railway coach. Every aircraft plant in the nation was open to his inspection and he was treated like a foreign prince wherever he went. But there were two events of that unique trip which he remembered with special clarity. One was his visit to the great Zeppelin plant at Lake Constance; the other was his meeting with a young and little-known German technician whose genius impressed him greatly. The name of that young man was Willy Messerschmitt.

Why the Germans should have been so anxious to impress McCurdy that year is a riddle he has never been able to solve.

The early development and fruition of such agencies as Trans-Canada Air Lines is partly an outgrowth of McCurdy's belief in the value of commercial air transport. It is a strange coincidence that Gordon R. McGregor, president of T.C.A. since 1945, began his professional career with the engineering department of the Bell Telephone Company and later saw highly distinguished service as a pilot in the Second World War under the command of some of the men trained at McCurdy's Toronto flying school.

McCurdy continued as president of Curtiss-Reid until September of 1939. Then, with the world again trembling on the brink of catastrophe, he promptly resigned his post and offered his services to the government. His boss now was Canada's iron-man minister, the Honourable C.D. Howe, and his official job went by the lengthy title of "Supervisor of Purchasing and Assistant Director-General of Aircraft Production."

His duties were as many and as varied as the title would indicate, and the skill and energy with which he discharged them made a superb contribution to the Canadian war effort for which he was later made a Member of the Order of the British Empire.

But the highest of McCurdy's honours was yet to come. One summer evening in 1948, while enjoying the cool green of his Ottawa lawn with some friends, a reporter telephoned. It was to congratulate him upon becoming the Lieutenant-Governor of Nova Scotia. The appointment took him completely by surprise. Why Prime Minister King should make the choice without first getting McCurdy's consent and why he should make the announcement to the press before he did to the man himself was never explained.

When besieged by the press for a comment, McCurdy suffered one of his rare moments of awkwardness and said that he appreciated the new honour and would perform his duties "as well as a country boy from Cape Breton could."

Seldom, if ever, has the appointment of a lieutenant-governor been greeted with so much publicity and favourable comment. There wasn't a journal in the nation which did not have an editorial about the event, and all of them applauded. But perhaps none of them spoke the thought of the nation better than the *Ottawa Journal* when it said:

"In our midst in recent days have been notable visitors—not the least among them, His Honor, the Lieutenant-Governor of Nova Scotia. J.A.D. McCurdy, minus top hat, uniform and sword, looks just as when in war days in a sports coat he sat in his office on Wellington Street and helped to get planes for us. He is not overwhelmed or made pompous by the new honor that has come to him. His clay, from Cape Breton, is different. Yet one feels, as he packs up his belongings here to betake himself permanently to Halifax, what a difference he will make to old, grey, stodgy Nova Scotia House, taking a fresh breeze into its cobwebby corners, giving it the grace of laughter, and also giving it class."

The McCurdys certainly lost no time in getting rid of the cobwebs in Nova Scotia's Government House. This great Georgian edifice is one of the oldest State residences in the Dominion, its cornerstone having been laid in 1800, and no house in all the new world has entertained more Royalty. Among the guests it has entertained have been Prince Edward, later Duke of Kent, and the father of Queen Victoria; Prince William Henry, later King William IV; Edward, Prince of Wales, later King Edward VII; George, Duke of Cornwall and York, later King George V; Edward, Prince of Wales, later King Edward VIII; Prince Arthur, Duke of Connaught; Princess Louise, wife of the Governor-General of Canada, the Marquis of Lorne; Prince Albert, later King George VI; Prince George, Duke of Kent; Prince Alfred, youngest son of Queen Victoria; Her Royal Highness, Princess Alice; and the present Queen and her consort, when she visited Canada as a Princess.

But time and the neglect of those not interested in tradition had allowed the grand old place to become badly run down. With his wife's good taste now serving as an admirable guide, McCurdy set about a programme of restoration. Partitions, some of them hastily made with beaverboard, which had been put in from time to time by more careless officials, were now removed to bring to light such things as ancient doorways and cupboards long since forgotten. Carvings, panellings and mouldings which had been buried deep under many coats of paint were cleaned and brought back to their old grandeur.

One of the loveliest features of the old sandstone mansion is a beautiful hanging stairway just off the front entrance. In all Canada, there is nothing like it. But when the McCurdys took over, they found its unique beauty was completely encased in an ugly closet.

Room by room, the historic old building was brought back to

its original glistening glory. Perhaps the most famous room of all is the Royal Bedchamber, which is a very large room on the second floor, with a canopied four-poster bed whose mattresses are piled so high off the floor that two steps had been built alongside to lead into it. When Edward, Prince of Wales (later King Edward VIII) was a guest here, so some Haligonians love to tell you, he looked in dismay at the ladder ascending to the heights, then yanked the top mattress to the floor and slept on it for the night.

But with the remodelling of the Government seat completed, McCurdy was not content merely to sit in it and look officially handsome. "Tradition and ceremony are apt to be dismal, or even ludicrous, if not done correctly," he declares. And in that belief, he worked hard at the task and with typical care and conscience.

He met the great of the nation and of the world, launched ships, decorated heroes, and cheerfully endured a thousand cultural functions. It was his job, of course, to open the Legislature, and to represent His Majesty in Nova Scotia. In that capacity he occupied the same desk which had been used by the most beloved Nova Scotian of all times, Joseph Howe.

And in addition to all these duties, there was always the endless job of speechmaking. Probably no Lieutenant-Governor in history ever made more speeches, for McCurdy was not only a very picturesque official, he was an excellent speaker as well. Nor did he confine his talks to recollections of his days as an early-bird. He had as many ideas as he had interests and he never hesitated to promote an unorthodox idea if he really believed in it. One of the things in which he always firmly believed was that there should be a union of all the English-speaking peoples of the world. That he had given the matter careful study is indicated by the detailed suggestions he once outlined to make such a union possible. Said he at a Canadian Exporters' Association luncheon in Toronto in November of 1949:

"The union authority should have somewhat the same limited but effective powers as set forth in the American Constitution, notably to make war and peace for all of its members; to govern their foreign relations and their non-self-governing territories; and to provide them with a common defence force, a common free trade market, a common currency, a common postal and communication system and a common citizenship; each country, however, preserving the present status of its institutions in so far as deemed practicable . . . .

"The difficulties which might be suggested were under full discussion by the Continental Congress at the time the union of the Thirteen Colonies was debated. The makers of the American Constitution ended by shifting only five major powers from the State governments to the Union government. They abolished thirteen independent armies; thirteen currencies; thirteen tariff walls; brushed aside thirteen barriers to communications; abolished the many dangers of being divided into Pennsylvanians, New Yorkers, Rhode Islanders and so on.

"It is now time, I think, that we call a convention to decide whether or not such an idea cannot be extended to include all English-speaking nations everywhere in the world."

On another occasion, his mind turned again to matters of science and he suggested that our medical men might very well explore the possibility of making spectroscopic analysis of human tissue and blood in an attempt to discover the cause of disease.

But whatever he did in his post as Lieutenant-Governor, it was McCurdy's talent for maintaining the common touch, in spite of the required dignity, that endeared him most to the thousands who came to know him.

When, in 1951, Nova Scotia received notice that the province would be honoured by a visit from Princess Elizabeth and the Duke of Edinburgh, the McCurdys immediately set about to provide a welcome that would be as unique as it would be proper. What should they serve such a distinguished couple at the State banquet? The following letter, which was posted in many of the stores and post offices back in McCurdy's home county, tells its own story.

"To my friends of Victoria County:

"Her Royal Highness the Princess Elizabeth and His Royal Highness the Duke of Edinburgh have graciously consented to dine with His Honour the Lieutenant-Governor and Mrs. McCurdy at Government House on the evening of November the 7th.

"My wife and I thought it would be nice and very appropriate to serve something as coming from my native county. I would be most grateful if my friends could make this possible.

"I am sure that nowhere else throughout Canada will their Royal Highnesses have an opportunity to taste a Cape Breton partridge. Our Cape Breton partridge season opens on October 15th, and as these birds cannot be purchased, my suggestion is that those of my friends who wish to do so, would leave with Lister MacLennan, Victoria Farmers' Co-operative, a partridge or two where they can be kept in cold storage till the appropriate date for shipment to Halifax.

"We require in all about 30. All contributions will be personally acknowledged, and I feel that the thanks of Government House will express to you all our appreciation of your generous act. Their Royal Highnesses will, I am sure, be delighted.

"J.A.D. McCurdy,
"Lieutenant-Governor"

He got the partridges.

For the information of those who are strangers to the laws of Nova Scotia, it should be explained that the partridge, a ruffed grouse, is a scarce delicacy which is rigidly protected. At that time, only fifteen could be shot by any one person in one season, and the bag was limited to five in any one day. Partridges could not be sold.

Asked afterwards what his impressions were of the young Princess, McCurdy thought her altogether lovely and clever, but just a little frightened. Asked what he had to tell her, he said: "I told her to stay just the way she was, and not to try too hard to grow great or serious. If she could do that, I told her, she would make a wonderful sovereign some day."

It was back in 1949 that this writer first met the Honourable J.A.D. McCurdy. It was in Halifax, on a lovely June day when that city was the scene of the annual convention of the Canadian Authors' Association. I had never been included in that august group before, and when the invitation to Government House came, my first reaction was to excuse myself as an altogether undistinguished and unnecessary addition to the press of greatness that was to parade itself that afternoon.

"But we are being received by the Lieutenant-Governor himself!" one of the good ladies explained. "Of course you will come! No, you won't have to wear a monkey suit!"

And it was while I was a part of the file which slowly worked its way toward the vice-regal handshake that I first began to realize how different Nova Scotia's McCurdy was from the traditional official dignitary.

"If he had been born in a hero-worshipping country like the United States," one of my writer friends remarked, "he would have been a Daniel Boone by now. But whether we talk about him or not, he's the greatest Canadian of our time."

When we finally reached the colourfully appointed reception room where McCurdy and his beautiful wife awaited us, there was certainly nothing about the man himself to indicate that he concurred in that high opinion. I can't remember now whether he had even taken the trouble to wear formal dress or not, but I do recall that in less time than it takes to tell, his boyish capacity for genuine interest and enthusiasm had changed that reception into a cosy little affair with all of the superior people as human as the rest of us.

I remember the way the table was decorated, too. Not with the wares from a florist's shop but with daisies and other wild flowers. Mrs. McCurdy confessed later that she had picked them herself.

I came to know Douglas McCurdy well. I treasured that friendship as one of the greatest privileges which has yet come my way. Indeed, what greater honour could come to a writer than to be permitted to know and to act as biographer for a man whom history may very well declare one day to be "the greatest Canadian of our time?"

Douglas McCurdy died in 1961. Days after a service in Montreal, the Royal Canadian Air Force, of which he was an honourary Air Commodore, with tremendous respect and ceremony, transported his ashes for burial in Cape Breton aboard a DC3-Dakota. On board was McCurdy's wife Margaret, son J.R.D. "Bobby," and his daughter Dianna. The pilot of the Dakota was Bob Fassold.

J.A.D. McCurdy is buried in a Baddeck cemetary overlooking the lake from which he flew the *Silver Dart* in 1909. A fly-past by R.C.A.F Sabre Jets paid tribute to the committal service.

Young Douglas McCurdy's Diary

The following are pages from the notebook which the young McCurdy wrote during the summer of 1898, the first summer when he and his two brothers lived in their home-made houseboat on Lake Bras d'Or. Spelling, punctuation, and other errors are his own. We have done nothing to alter it except to divide it into paragraphs and thus force his tireless prose to take a few restful breaths.

We live in a cottage on the shores of the Brass Dor Lake There are four children in our family, one girl and three boys. Susie is 16 George is 14 Douglas is 12 and Lucian is 10 years old. Four years ago we started to build a houseboat but since then it has grown to be quite a habitation and we have lots of fun in it. It is built in this way. There are eight water tight casks laid on their sides and boxed in with a frame and a platform 19 ft long and 10 ft wide built on top. Three ft from the edge of the platform is a frame with the tent on it over the tent is a fly which acts as a double roof. It does not come down to the edge of the platform but is tied down with codline put through eyes. We came over to live in the houseboat on July 11th 1898. We landed in the boat on the opposite side of the bay which we named "Point Look Out" We soon transfered our baggage to the houseboat and proceeded to tow her down to Calm Harbour. When we got about half way down George and I were rowing in the boat with the house boat in tow we took it into our heads to jump overboard and have a swim so we got undressed very silently and all at once we jumped into the water and made a big splash. Lucian and Danny McInnis (a friend of ours who was with us) came running to the door to see what made the splash. Before we got dressed we saw a boat load of people coming down to see us. They turned out to be Papa Aunt Hannah Aunt Georgia and Mrs. Kennan. When we got a little bit past "Beloved" a easterly breeze came up so we could not tow her against the wind so we took her in near to the shore and then we got undressed and towed the house boat with our hands. When we got her fastened in the harbour by means of a rope Danny went home and George and I went for water soon we had supper and went to bed. When we woke up this morning George and I went home to get a supply of provisions. When we came back Lucian had breakfast all ready for us Our breakfast consisted of oatmeal, milk, bread and butter and tea. After breakfast we towed the house boat up to "Point Look Out" and secured it to a stick then we went in for a swim. Just before dinner George went home to get Susie who intended to stay all night. We had dinner and with susie's help we soon had the place all in order Susie then went for a row and we went in for a swim. After Susie returned I went home with her. Before we returned Jamie Atwater a very nice friend of ours came over from town and helped George and Lucian to tow the house boat down to "Elsie Harbour" When we came back we halled the boat accross the beach and went onboard the house boat and after we had supper we turned in for the night. It is a very fine day. George you make the breakfast Susie Lucian and I will go home after breakfast. When we got home we found that we were going to go to Middle River to fish so Susie and I came back and told George We soon came home and got started. We came back after dinner on July 15. The first thing we did was to tow the houseboat down to "Point Look Out" We staid down their a good while when we thought we would like to have it up to the "Dump" So we towed it up. And George and I went in for a swim After we got dressed we staid inside for a while and read. At six o'clock we started to get the supper. We got all through besides making the bed at 9 o'clock. The way we made the bed was:— We had a bedstead or a sort of a lounge with a straw matrass on it with the bed all ready to sleep in, on top of the other bed was another matress which was not so large as the other one so we pulled the edge of the upper matrass over the edge of the lower matrass and folded up the bed cloths belonging to the upper matrass and put them behind the upper matrass so as to make it all leval Then we put one quilt over the whole business. So at night we just put three camp stools opposite the bed and put the upper matrass on then this made it into one bed so the three of us slept in a bed 6 ft wide and 6 ft long. We read untile ten o'clock then we put out the light and went to sleep. When we woke up this morning it was raining very hard so we did not get up untile 0 o'clock. When we got dressed we had breakfast. We then went up to the landing to get a flag pole. We got a poplar and stripped off all the bark and brought it down to the "Dump" and put the flag up on it. When we were done we had dinner and sailed around for a while landing at various places we had supper at four o'clock after which we went home.

We came over Monday after dinner on July 18 We always went home on Saturday night so as to be ready to go to church on Sunday A little while after we came over Miss Lena McCurdy came down to the "Dump" and said that Grandpa wanted the keys which belonged to the house up in town, so I had to go home and get them. I did not come back until half past seven so we soon had supper Papa came over to stay all night in the house boat. We read a little while then we went to bed. While we were in bed we thought that we would like to go cod fishing so we set the alarm at half-past two o'clock we soon went to sleep. Lucian sleeping with Papa and George and I sleeping together in the kitchen, George was the first to wake so he woke me and we made the porridge and put on the tea. Lucian soon woke and got dressed so he and I went up to the landing to get the anchor and rope that was anchored in the harbour we soon came back and had breakfast When Pape got up he went fishing in the harbour. George and Lucian got some clams and we started fished a good while but did not get anything. Then we thought we would like to go to "Spectacle Island" so we went to Mr. Bell's house and telephoned to Papa and he said we could go. Maggie Campbell Mr. Bells cook gave us a lunch. When we got to the "Island" we got some bull rushes and Lucian got two sand pipers We ate our lunch and started from the Island for Baddeck We had fair wind and a good breeze so we soon got there We then sailed to the Ware House and from there we went home We came back in the after noon and had our supper after which we went to bed. When we woke up in the morning it was raining cats and dogs We did not get up until 9 o'clock After we had breakfast we started for home in the sail boat It was blowing a heavy gale out side the harbour We tacked up as far as Mr. Carruth's and we could not go any further so we landed on the shore and walked up to the cottage. It was very wet. When we got back we changed our clothes and stayed indoors all the after noon We had a very nice supper and we went to bed at 9 o'clock.

It is raining still We got up about 10 o'clock After we had breakfast George went home. And Lucian and I went for water up to the falls.

We had to go in a old dory because George had the boat. The falls were very pretty because there had been such a rain storm. When we got back I made pan cakes for dinner just then George returned and we had our dinner After we had washed the dishes George read a book Lucian wrote his diary and I fooled around until 7 o'clock when Lucian was done Then I began to write and am writing until bedtime which is twenty five minutes past ten o'clock.

When we woke up this morning it was about ten minutes past 9 o'clock Before we got dressed George got up out of bed and went into the kitchen and lit the lamp so as to have the breakfast done by the time we were ready for it. I read my diary over to the boys before we got up Just as I was done we herd a boat comming down the harbour We jumped out of bed in a hurry to see who it was. We found that it was Captain Frazer, 2 boys and one girl We thought they were comming down here so we got dressed in a hurry but they did not come While we were having breakfast Mary McLeod and her sister Annie came over with some milk. George then took them back and Lucian and I washed the dishes. We used to put the bottles of milk over board so as to keep it cool but sometimes it got sour so Lucian went over to a spring to make a dam to keep them in the spring is so cold that at night it sometimes frizes. I am left all alone writing my diary. I forgot to say that before we got up we had some excitement over a mouse. I was awakened by something running over my face Just as I opened my eyes I saw a mouse running down the back of the bed The boys soon got up and we all looked for it We soon found it hiding in my shoe as it ran out Lucian cut off its head with the hatchet and threw it into the lack. After I wrote my diary I read a book called the "Standard Bearer" Lucian came to say that he finished the place for the milk so I went over to see it. When we came back to the house boat George returned with somethings to eat also a valise with some clothes for us to wear down to the Point. George then went over to the spring to get the milk while he was done the dory went adrift so I had to strip off and swim after it Lucian soon got the dinner. When we were just about half through John McAuly and Walter Hannan came over from town to dig clams We thought that we would like to go codfishing too so we got a basket full of clams and had things all ready to go. A little while after wards Jamie came down so we went aboard the houseboat and cleared up the dinner things Jamie read us stories. Soon we had our supper then we went up to the ware house in Jamies canoe and walked up to the Point. We had a very nice time and returned at half past ten o'clock. We set the alarm at half past two o'clock. The alarm soon woke us up but we were too sleepy to get up until four o'clock As soon as we got our breakfast we started for the banks which were just off the Point. As Jamie and I were in the canoe we got the first we got the line and hook all ready when to our disgust we found that the boy's had all the clams so of course we had to wait until the boys came. We got a few clams from them and Jamie bated the hook. He caught the first one and it was a good size I tried next I soon felt a bite and I halled in the line Yes there was one sure enough caught in a very queer way. It had been swimming around the hook and I jigged it. By this time it was Lucian's turn he soon caught a pretty big one then George tried. He was not so successful as the rest of us. He only got a few perch. After Jamie was done fishing he started for home so we went up to the Point to breakfast as Mrs Bell had invited us the night before. We gave the biggest fish to Mrs Bell and she was very much pleased with it. We got home about half past twelve Just as we were coming down the harbour we saw a whole partie comming over to see us in a sailboat so we hurried as fast as we could so as to be there first as we did not make the bed in the morning The partie consisted of Louise and Mrs Blanchard Mr. Bouman Aunt George Mr and Mrs Munn. Aunt Georgie brought over some potatoes and as Louise was handing them to Jamie they fell into the water Jamie and I then went home with them after they had seen around the house boat also the falls which were very pretty, and brought back some milk Jamie and Lucian made the dinner and when they got about half through Papa came down in his sailboat bringing Mrs Bell and Mrs Kennan. We were in for a swim when they came so they sailed around until we were ready to receive them As soon as we cleared up they came abord and as Mrs Bell did not see the houseboat before she thought it was grand After a while they went away and as we were short of milk George went up to John McDermids to get some for supper, After we washed the dishes

Jamie tried to go to sleep George soon returned Lucian and took the canoe and went for a paddle As I was collecting flowers we went over to the "Dump" to look for some. When we came back George wanted to go over to the falls to get a drink so Lucian and I paddled him over, Jamie, Lucian and I made the supper. After we were through George and Lucian washed the dishes and Jamie and I loafed on the bed When the boys were done Jamie told us a story. After a little while we thought it was time to go home as it was Saturday night so we paddled up to the Lodge Wharf. George and Lucian went home in the boat and Jamie and I went home in the canoe. We did not come over to the house boat until 5 minutes past nine o'clock We made the bed in a very short time and George and I wrote our diary's. We did not go to bed until Jamie came which he soon did then we went to bed.

We opened our eyes at half past seven in the morning as we did not feel like getting dressed we stayed in bed until 9 o'clock. As it was our custom to go in for a swim before break fast we did so Lucian got the breakfast Jamie went for water George made a wharf from the house boat to the shore I cleared up. When Jamie returned we had breakfast I am Know sitting at the table writing my diary and George is trying to take my picture After George was done Lucian tried while I was in the same position In a little while George and Lucian went up the harbour to cork up a dory that Mrs Kennan gave us While they were gone I wanted to go a shore as I could not step ashore I took the broom to try and jump but I fell in the water When the boys returned it was raining As we were out of milk Jamie started to go for some. As he did not want to get his clothes all wet he put in George's trousers and one of his stockings and one of his belt Then he went ashore in the boat and got a coat of his out of his canoe He then started off While he was gone I read a story to the boys When Jamie came back we got the dinner. George and Lucian fooled around tile Jamie and I wore done washing the dishes We still had to go on another trip We were out of eggs so I got ready and started When I came back Papa and Miss McCurdy were in the houseboat I got aboard very silently and nobody knew that I was abord except George I peeked in the door and at last I came in. Miss McCurdy read my diary and Papa talked about various things.

After they went we had supper George and Lucian washed the dishes and Jamie and I lay down and talked because we had done all the work and were tired. The little flies were awful so we hung a candle lantern outside to attract them We went to bed at 10 o'clock We woke up in the morning about 8 o'clock. George and Lucian went home Jamie made the breakfast and I house cleaned When the boys came back we had our break fast In a little whill we went up to the Point.

We left Jamie up there and got back about half past 2 o'clock When we started to make the dinner we found that there was no oil in the lamp so we made a meal on cold ham and bread and butter and milk After dinner we went in for a swim We took a lot of pictures Among some I took one of George and Lucian on the dory just going to jump in After we were done swimming George went for a sail and Lucian was trying to right the dory and we fell in the Lack he got ashore and took off his clothes and put them on the bank and put on dry ones and then he read the "St Nicholas". I took Jamie's canoe and paddled around the point to see where George was I saw the sail up at the wharf so I paddled back and told Lucian When George came back Lucian and I were down at "Point Look Out" with Mr. Frost one of our friends so as the boat came past Lucian jumped in and went with George I went back in the dory as I had left it in the shore. After we had put on our clothes we went up to the "Point" to supper as Mrs Bell had invited us. Mr. Bell came back with us as he wanted to see the houseboat George took him back to the "Dump" after he had seen all and he went back to the point When George returned we had out usual bedsupper then we went to bed We woke up about half past seven in the morning George and I made the bed and Lucian went in the boat after the dory which had gone adrift during the night When he came back we went in for our usual swim after which we had breakfast Just as we were done Rory McLeod came over to say that Grandpa wanted us to rake up at the house in town in the afternoon. George took a photograph of him in the boat After he went we went in for another swim and stayed in until quarter past one We soon went home and went up town on our byciles When we got up as far as Henry Tailors we saw the Harlow going out with Papa Mr & Mrs. Bell, Daisy Bell Susie and Charles Thompson on board. They were going to Sable Island We stopped and waved our handkerchief at them but they did not see us. We came back at six o'clock bringing over with us some grub and soon had supper a little while after we went to bed. When we woke up this morning we went in for a swim George and Lucian got dressed and I made the breakfast Just before we had breakfast we saw a boat coming with Uncle Henry David Maggie Lottie Aunt Georgie and Aunt Hannah in it so I got dressed as fast as I could The girls all went away and we went in for a swim with Uncle Henry and David and had a lovely time. Of course we got dressed first and had our breakfast over by the time that they were ready. We sailed them home in our boat. As it was Grandpa's birthday we stayed home all the afternoon In the evening I went down to Port Beves with Clark McCurdy and did not return until August 3rd.

When I was coming up from the gait I met Aunt Georgie just going down to the wharf to blow the horn for the boys as Grandpa wanted them to rake hay up in town as they did not come I had to go. I came back about six o'clock and after supper I went over to the other side with Miss McCurdy. I walked down to the end of "Point Look Out" and called to the boys. They were over at "Beloved" so they came over and took me back. Just as I got there we went in for a swim by lamp light Soon after we went to bed. When we woke up this morning George made the break fast and Lucian and I went for a swim Just as we got abord the house boat Roland Ellis came down to ask George if he would take part in a play and of course he excepted After breakfast we went home to rake hay We brought Clark back with us to stay all night Clark and Lucian slept together and George and I slept together in the kitchen About 2 o'clock in the night Jamie called in to see us but of course we did not remember it until we were reminded. We got up about 8 o'clock and there was an awful gail blowing After breakfast we went home I was going to have a birthday partie so I did not go back I went up town to send around invitations The boys came over in the afternoon We had a very nice time I had Miss McCurdy, Jamie, Fairy and Baby Campbell, Robert and Roddie Bethune, Earnest Hart, Daisy Bell, Eulah Tremain, Mrs. Bell, Clark Bell (and our own family) While we were playing a game Papa took a photograph of us as you see it here. We did not go back that night as it was very late We did not come back until the 10th. George and I came over today and brought the house boat up from "Beloved" to "Point Look Out" We started to clean the house boat so we took all the things out and put them on the shore. Before we were through Johnny McNeil came over and we went in for a swim with him and had a very nice time. When we got through cleaning we had dinner After dinner we went down to "Baby Harbours" to get some flowers when we came back it was pretty late so we had our supper then went to bed and read we soon went to sleep. When we woke up in the morning we had break fast as George wanted to read I went home and was to bring his dinner back I had a good dinner at home and sailed two Indians over to the Lodge Wharf and then came back to the houseboat After George had had his dinner he went home and I wrote my diary and read a book Soon George came back and we started to go down to Beloved to get some nails but it started to rain so we came back As we were talking we herd somebody calling us so we looked out and their was Susie and Susie Stairs on the other side so we went over and got them they stayed for about an hour then we took them home. We soon had supper and then went to bed.

When we woke up this morning it was blowing very hard I went out to get the milk but found that it had rolled down the bank so we had to eat our pirrage without any While we were looking for it the stick that the house boat was tied to gave way and as the wind was blowing easterly the houseboat began to drift very fast so I jumped over board with the roap and swam ahead of the house and pulled her on the shore then we let out more scope and I tied her to a stick on the other side of the beech but as it was blowing very hard we took it up to Falls Harbour and tied it just out side. We soon had supper and went to bed. We woke up on a beautiful mornmg.

We went home but as we thought that there was not enough sail on the boat we made 2 more gibs and a topsail as we were going back we meat Jamie and Roland comming down in the canoe they caught up to us and we went to the house boat and had a swim Roland did not swim at all but just splashed the house boat They staid to supper and it was blowing "great guns" out side so they told us not to sail home when we started to go home we put the house boat in order and started to tack up the harbour and when we got up as far as the "landing" the mast broke we landed and nailed it together and started again but it broke again so we turned and went back and staid all night at the houseboat. It was Sunday when we woke up so we got our breakfast and rowed home with a broken oar. We returned on the 17th of August We started with Papa's boat and sailed over to the "Lodge Wharf" and got our boat in tow and started down to the houseboat When we got pretty near the "Dump" we could not stear the boat off the shore so George let our boat loose and we went along to the houseboat We came over to the house boat and I got off then George went over to the other side to get the boat he said that the boat was away out in the Lake so he took off all his clothes and swam out for it we got the house boat all ready and Neil McDermick met us at the "Dump" with a horse old Hary to hall the house boat up to the "Rustic Bridge" so we fastened the roap to a cleat and started the horse. George and I were keeping it off the shore with oars but all of a sudden the cleat gave way so we tied the roap to the corner of the fly frame but it gave way so we took the fly off and put it in side. This time we put the roap around the floor so this time it held the horse pulled it up to the landing but as the bottom was too soft Neil took the horse around and George and I towed it by hand around the point past the landing Here we hitched the horse on again and he pulled it up to the Rustic Bridge Here we tied it. We took a load home in Papa's boat We came over in the morning and took the house boat frame all to peaces Neil came down with the horse to hall it up but the old horse could not so we left it there Lucian and I went home in the Laboratory boat and George went home in our boat We did not try to hall up the houseboat until September 17th When we got there 2 men met us with the 2 farm horses they just walked it up We then took out all the casks and put them in Mr Bells Boat House The house boat is now gone to "sleep" for the winter not to be used untill next summer when we hope to have better "adventures and perils"

Douglas McCurdy
Read aloud by Douglas,
December 7, 1898.

The Other Members of the Aerial Experiment Association

Alexander Graham Bell
LL.D., Ph.D., D.Sc., M.D.

Alexander Graham Bell was born in Edinburgh, Scotland, on March 3, 1847, the son of Alexander Melville Bell and the former Eliza Grace Symonds. He was educated at Royal High School, Edinburgh, Edinburgh University, and University College, London. He came to Canada in 1870 and settled at Brantford, Ontario. He became professor of vocal physiology at Boston University and patented his invention of the telephone in 1876.

In collaboration with his cousin, Chichester Bell, he developed the method of recording on wax. This invention was later sold to Thomas Alva Edison. He was awarded the first patent for the telephone in 1876. In 1877 he married Mabel Gardiner Hubbard and took up residence in Washington, D.C. With their daughters, Elsie May Bell and Marian Hubbard "Daisy" Bell, Dr. and Mrs. Bell visited Baddeck, Cape Breton, and established a summer home there in 1886.

At Baddeck Dr. Bell carried on extensive experiments, many of them in connection with heavier-than-air flight. From these experiments the Aerial Experiment Association was developed at the suggestion of Mrs. Bell who, incidentally, had been deaf since the age of four. Dr. Bell, a former teacher of the deaf, was greatly interested in problems relating to deafness, and in 1887 he founded and endowed the Volta Bureau in Washington for the increase and diffusion of knowledge relating to the deaf. In 1909, at Dr. Bell's suggestion, this institution was presented to the American Association to Promote the Teaching of Speech to the Deaf, an organization which he had also founded and endowed. He continued a wide range of experiments at Baddeck, including development of hydrofoil watercraft. His HD-4 at one time held the world speed record.

He was appointed the Regent of the Smithsonian Institution in 1893, and received the French Government *Volta Prix* of 50,000 francs in 1881 for the invention of the telephone. He was a recipient of the Albert Medal of the London Society of Fine Arts in 1902, the Hughes Medal of the Royal Society in 1913 and the freedom of the city of Edinburgh in 1920. He was a member of the National Academy of Sciences of the United States, and one of the founding members and one time president of the National Geographic Society.

In 1922 Bell died at his summer home, Beinn Bhreagh, in Cape Breton, and is buried there.

Frederick Walker Baldwin
M.E., D.Eng.

"Casey" Baldwin was born in Toronto, Ontario, on January 2, 1882, the son of Robert and Elizabeth Mary Baldwin and grandson of Sir Robert Baldwin, a Prime Minister of Canada before Confederation. He was educated at Bishop Ridley College, St. Catharines, Ontario, and the University of Toronto, from which he was graduated in mechanical and electrical engineering in 1906. He married Kathleen Stewart Parmenter of Gananoque, Ontario, on June 12, 1908. He was chief engineer of the Aerial Experiment Association from 1906 to 1909, and continued to work as an engineer at Dr. Bell's laboratories from 1909 to 1932. At Hammondsport, New York, on March 12, 1908, he was the first British subject to fly an aeroplane. He gave up flying in 1911. He was elected a member of the Legislature of Nova Scotia for Victoria in the general election of 1933 and president of the Nova Scotia Conservative Association in 1937. He received the honorary degree of Doctor of Engineering from the Nova Scotia Technical College in 1946. He died at Neareagh, Nova Scotia, in August, 1948.

Glenn Hammond Curtiss
D.Sc.

Glenn Hammond Curtiss was born in 1878 in Hammondsport, New York. He established a motorcycle factory there in 1902. Between 1905 and 1907 he set motorcycle speed records. He designed the engine for the first United States military airship in 1905, and continued to design motors for aircraft and dirigibles from 1907 to 1909. Curtiss formed the first aircraft manufacturing company in the United States in partnership with A.M. Herring.

He won the *Scientific American* trophy for the first public aeroplane flight of a mile in the United States in 1908, and in 1909 he piloted his *Golden Flyer* at Rheims, to win the Gordon Bennett International Trophy for the fastest speed over a twenty-kilometre course. He won the *New York World* prize for a flight from Albany to New York in 1910, and later that year persuaded the American Navy to fit a wooden platform over the deck of the cruiser *Birmingham* for a test flight. Eugene Ely made a successful flight from this platform on November 14, 1910, in a standard Curtiss biplane at Hampton Roads, Virginia. This was the beginning of the aircraft carrier. Curtiss built and, on January 26, 1911, flew the first really practical seaplane. He built a hydroplane in 1911 and a flying boat in 1912.

Curtiss established aviation schools. His aircraft factory supplied machines for the United States, British and Russian forces during the First World War, and developed flying boats and other types of sea and air craft for the United States Government. Considered the greatest American aviation pioneer and inventor after the Wright brothers, Glenn Curtiss died at Buffalo, New York, in 1930.

Lieutenant Thomas E. Selfridge

Lieutenant Thomas E. Selfridge was born about 1882 in San Francisco. He was a nephew of Rear Admiral Thomas Oliver Selfridge of the United States Navy, and was graduated from the United States Military Academy, West Point. He is considered the seventh man to have flown a powered heavier-than-air craft. He flew the *White Wing* on May 19, 1908, at Hammondsport, New York. On September 17, 1908, he was passenger and representative of the United States Army in a test flight by Orville Wright at Fort Myer, Virginia, near Washington. The Wright plane crashed, severely injuring its pilot and Selfridge, who died in Walter Reed Hospital, Washington, a few hours later—the first person killed in an airplane crash.

Selfridge Field [today's Selfridge Air National Guard Base], the United States Army Air Force base near Detroit, is named in his memory.

Last Notes

Author of *The Silver Dart,* H. Gordon Green was born in 1912. A graduate in science and arts of the University of Michigan, he served in both the Canadian and United States armies during the Second World War. He is best known to Canadian readers as a writer of short stories and for his novel, *The Praying Mantis,* which received a Hopwood Major Award for fiction in 1949. He wrote for television and motion pictures. He lived on a farm near Ormstown, Quebec, where he pioneered in the breeding of new and unusual types of livestock. Mr. Green was secretary of the Canadian Landrace Association and of three other national breed organizations. He was the last general editor of the *Family Herald* and in 1958 he was elected president of the Canadian Authors Association.

Our thanks for help with this new edition of *The Silver Dart* to Chip Bird, Field Unit Superintendent for Parks Canada, and to Valerie Mason and Madeline Harvey of the Alexander Graham Bell National Historic Site. Parks Canada supplied digital copies of most of the photographs in this book. Photos of the replica of the *Silver Dart* on the ice, and wrecked, in 1959 are courtesy of James Fader. Photos of the two 1984 *Silver Darts* courtesy Gordon and Laverne MacRae; and of the 2009 Centennial reconstruction and flight courtesy Doug Jermyn, president, AEA2005, and Eleanor Anderson, Project Manager of the 2009 Silver Dart Centennial Committee. Photographs of the 2009 flight in the book, and of members of the AEA2005 team that built that plane, are by Vaughan Merchant. Front Cover: Marian Bell Whitcomb.

For the original edition of *The Silver Dart,* the photo of Alexander Graham Bell and staff hauling down a kite and of Melville Grosvenor hanging from a kite rope were taken by Gilbert Grosvenor; the interior of "The Cygnet" kite was photographed by J.A.D. McCurdy; the glider at Hammondsport, the crash of the *Red Wing,* the five members of the Aerial Experiment Association by Mabel Bell, as well as the visit to Hammondsport by balloonist Augustus Post, were all courtesy of the Bell family and National Geographic Society.

Last Notes

The H.M. Brenner photographs of the February 23, 1909, flight of the *Silver Dart* and of skaters pushing the *Silver Dart* on February 24th, and the John MacNeil photograph of the *Baddeck No. 2*, were courtesy of the National Geographic Society. McCurdy at the controls of the *Silver Dart* in Hammondsport and the Curtiss engine on the ice boat, courtesy of the Smithsonian Institution. Maurice Crosby took the photograph of the Hon. J.A.D. McCurdy as Lieutenant-Governor of Nova Scotia greeting Capt. D.L. Burleson of the Royal Canadian Army Service Corps, courtesy of the *Halifax Herald and Mail*. The three McCurdy brothers working on the wing of an aircraft, and other family photographs, were courtesy of the Hon. J.A.D. McCurdy. The sod-turning ceremony was courtesy the Volta Bureau and the National Geographic Society. McCurdy holding a small model of the *Silver Dart* was courtesy of Canadair Limited, and the photograph of the full-sized replica of the *Silver Dart* at Trenton, Ontario, was courtesy of the R.C.A.F.

There are several full-scale replicas of the *Silver Dart* on display, including at the Atlantic Canada Aviation Museum near the Robert L. Stanfield International Airport, Halifax; the Reynolds-Alberta Museum in Wetaskiwin, Alberta; the Canada Aviation and Space Museum in Ottawa; the Royal Air Force Museum in Hendon, England; and the National Air Force Museum of Canada in Trenton. The second plane built by the R.C.A.F.—a non-flying reconstruction—is in the Glenn H. Curtiss Museum in Hammondsport, New York. It is not on display. The replica built by the AEA2005 team and flown in 2009 has found a permanent home at the Alexander Graham Bell National Historic Site of Canada in Baddeck, Cape Breton, Nova Scotia.

There is a display of photographs honouring J.A.D. McCurdy, the Aerial Experiment Association and the flight of the *Silver Dart*, as well as a bust of McCurdy, at the J.A. Douglas McCurdy Sydney Airport.

There are several videos about the original 1909 flight of the *Silver Dart* as well as the 2009 Centennial Flight at www.youtube.com. The voice of J.A.D. McCurdy remembering his 1909 flight can be heard at: http://thumbnails.cbc.ca/maven_legacy/thumbnails/td_silverdart19490223er1__828629.mp3

God's Country

17 CAPE BRETON STORIES—CLASSIC AND RARE

selected with an introduction by Ronald Caplan

AS CAPE BRETON ISLAND continues to earn its place at the table of Canadian literature, *God's Country* offers an essential collection of classic stories that helped build that award-winning reputation, as well as several rare and harder-to-find stories that point toward future literary achievements.

Under one rich and attractive roof: Alistair MacLeod, Joan Clark, Lynn Coady, D.R. MacDonald and many more.

224 PAGES • $18.95 • ISBN 978-1-926908-12-0

As True As I'm Sittin' Here

200 CAPE BRETON STORIES

edited by Brian Sutcliffe & Ronald Caplan

COLLECTED BY ARCHIE NEIL CHISHOLM—laughs, comebacks, ghosts, fairies, put-downs, and all-round wit—from Dan Angus Beaton, Jim St. Clair, Sid Timmons, Hector Doink MacDonald, Annie the Tailor MacPhee, and many more.

Stories that shorten the road, lighten the work, and fill the pauses between tunes, throughout Cape Breton Island.

220 PAGES • $17.95 • ISBN 1-895415-58-6

10 Nights Without Sleep

CAPE BRETON CELTIC COLOURS INTERNATIONAL FESTIVAL

by Dave Mahalik

AN INSIDER'S ACCOUNT of the music festival that brings the best of Cape Breton and international Celtic musicians and their fans to October's Cape Breton Island, for unforgettable performances, workshops and local culinary events — and the after hours Festival Club that is both the heart of the festival and the talk of the world.

With over 125 photographs by Murdock Smith.

156 PAGES • $18.95 • ISBN 978-1-926908-90-8

Breton Books

Wreck Cove, Cape Breton, Nova Scotia B0C 1H0

books@capebretonbooks.com • 1-800-565-5140

www.capebretonbooks.com